The Vanishing Newspaper

EW YO

ing
ers
sing

The Vanishing Newspaper

Saving Journalism in the Information Age

PHILIP MEYER

University of Missouri Press   Columbia and London

Library of Congress Cataloging-in-Publication Data

Meyer, Philip.
   The vanishing newspaper : saving journalism in the information age / Philip Meyer.
        p. cm.
   Includes bibliographical references and index.
   ISBN 0-8262-1561-0 (alk. paper) — ISBN 0-8262-1568-8 (alk. paper)
  1.   Journalism—United States. 2.   Journalism—Economic aspects—United
States. 3.   Newspapers. I. Title.
   PN4867.2.M48 2004
   071'.3—dc22

                                                                    2004016121

Designer: Kristie Lee
Typesetter: Crane Composition, Inc.
Printer and binder: Thomson-Shore, Inc.
Typeface: Berkeley

In memory of Edwin A. Lahey, 1902–1969

*"All I require of my publisher
is that he remain solvent."*

Contents

The Vanishing Newspaper

Introduction

J O U R N A L I S M is in trouble. This book is an attempt to do something about it.

The idea was born on Flattop Mountain in North Carolina in the summer of 2001. I was reading *The Sum of Our Discontent: Why Numbers Make Us Irrational* by David Boyle and following an Internet discussion about newspaper layoffs when I saw the possibility of making a connection between measuring quality in journalism and investors' decisions. By chance, the owner's library in our rented vacation house contained the Louis Lyons volume with the anecdote that opens chapter 1. That might have been the spark.

My original model was simple. Give investors numbers that would lead to better and longer-term predictions than the quarter-to-quarter earnings changes that they like so much, and they would encourage managers to focus more on the long-term health of their news organizations. If the market is efficient, better news products and community service would become another means of putting a value on newspaper companies.

There is no shortage of historical studies showing a correlation between quality journalism and business success, as Esther Thorson of the University of Missouri demonstrated when she reviewed the literature.[1] But there are two problems. One is that it is very difficult to show that quality journalism is the cause of business success rather than its by-product. In my lifetime, I have read many bad newspapers that made money and seen how the character of the owner could be the critical factor in determining quality. But even if we could gather enough data to produce the quality-success equation in a way that would convince investors, we would still face the other problem. We are not in a steady-state universe. The business model for news has been so disrupted by

---

1. Posted at http://www.unc.edu/~pmeyer/Quality_Project/quality_resources.html

1

new technology that the formula for success could be changing in un-expected ways. The past is not always prologue.

Good journalism has managed to survive, if not always to prevail, through many changes in technology in the past century. The Internet is just the latest in a long series of advances that contribute to the de-massification of the media. Richard Maisel, a sociologist, documented this trend a generation ago when there was no Internet and personal computers were still expensive and rare. He saw that the mass media system was contracting relative to the rest of the economy, and that spe-cialized media were expanding to fill the gap. Maisel found it happen-ing across a broad array of media forms. The number of off-Broadway performances, which are held in smaller theaters, was increasing, while the less specialized Broadway performances were declining. Newly built movie theaters were being designed with fewer seats. The field of technical books had more growth than fiction books. Bimonthly and quarterly magazines were doing better than monthly and weekly maga-zines. Community newspapers were holding on to readers more effec-tively than were metropolitan papers. National advertising in newspapers was growing less than local retail advertising while the most specialized form, advertising in the classified pages, was growing most of all.[2]

The Internet is accelerating this trend toward smaller audiences by giving seekers of specialized information an increasingly efficient source. Why would you look up yesterday's closing price of your favorite stock in the newspaper when you can find the last half-hour's price on the Internet? The umbrella newspaper has owed much of its success to its ability to provide a mosaic of many such specialized interests, but it is no longer the most efficient way to appeal to those interests.

The disruption of existing business models by substitute technology is an old story in American business, and a substantial literature has de-veloped around the problem. The old businesses hang on too long to their accustomed ways of doing things and become ripe targets for up-start competitors who are not burdened by tradition. The issue of quality journalism and business success has to be considered in this context.

---

2. Richard Maisel, "The Decline of Mass Media," *Public Opinion Quarterly* 37:2 (Sum-mer 1973): 159–70. Maisel's first version of this paper was presented at the 1966 an-nual meeting of the American Association for Public Opinion Research in Swampscott, Mass.

At the outset, I hoped to produce evidence that a given dollar invest-ment in news quality would yield a predictable dollar return that would more than justify the outlay. That might be possible, and the evidence in this book provides some support for the idea, but at nowhere near the level of precision that would excite an investor.

The main value of the work that follows is that it offers a model for looking at the news business that supports our intuitive appreciation for quality and, most importantly, can be transferred to whatever strange forms of media will convey the news in the future. The most interesting of these new forms are being invented by nonjournalists, and often they are ignorant of the culture of truth-telling and fairness that enabled the best news givers to prevail. That's not an insurmountable problem as long as there is enough varied experimentation going on to allow truth and fairness to emerge. Natural selection will do the job. Maybe we can help it along.

This book is an attempt to isolate and describe the factors that made journalism work as a business in the past and that might also make it work with the changing technologies of the present and future. The first two chapters present the theoretical model for journalism as a business that can be both advertiser-supported and socially responsible. Chapter 3 shows how advertisers are responding to new technology. Chapters 4 through 9 report the results of my quest for evidence that quality in journalism is good business. Chapter 10 recounts the recent history of increasing investor influence on newspaper companies. Chapter 11 considers the problem of harnessing the new and disruptive technolo-gies to the old values. And the final chapter is an appeal for solidarity among the men and women who do the day-to-day work of journalism and on whom the maintenance of its standards ultimately depends.

1

The Influence Model

O N C E when William Allen White visited Boston, two young reporters, Louis Lyons of *The Boston Globe* and Charles Morton of the *Transcript*, sought him out for an interview. The Sage of Emporia put his arms around the two men and said, "We all have the same face. It is not an acquisitive face."[1]

Later, when he was curator of the Nieman Foundation (1939–1964), Lyons recalled that moment and advised young journalists to work for organizations that put service to society above the values of "a banker or an industrialist." Today, he might say above the values of "an accountant or a stock analyst." But it's the same idea.

The glory of the newspaper business in the United States used to be its ability to match its success as a business with self-conscious attention to its social service mission. Both functions are threatened today. Measured by household penetration (average daily circulation as a

A version of this chapter appeared in *Newspaper Research Journal* 25:1 (Winter 2004), 66.

1. Louis M. Lyons, *Reporting the News* (Cambridge, Mass.: Belknap Press of Harvard University Press, 1965), 255.

percent of households), this mature industry peaked early in the 1920s at 130. By 2001, newspaper household penetration was down to 54 percent.[2] But while household penetration declined, newspaper influence and profitability remained robust. Now both are in peril.

The decay of newspaper journalism creates problems not just for the business but also for society. One problem is basic: to make democracy work, citizens need information. "Knowledge will forever govern ignorance," warned James Madison, "and a people who mean to be their own governors must arm themselves with the power which knowledge gives."[3] Democracy was more manageable when the mass media and their associated advertising for mass-produced goods tended to mold us into one culture. But that started to change after World War II. For some time now, historians have seen the world in three stages: a preindustrial period when social life was local and small in scale; the industrial period, which made both mass communication and mass production possible; and the third or post-industrial stage, which shifted economic activity from manufacturing to services. Journalism professors John Merrill and Ralph Lowenstein described the effect of these changes on our media system in 1971.[4] The mass media were already starting to break up the audience into smaller and smaller segments, promoting what sociologist Richard Maisel called "cultural differentiation." If we're all attending to different messages, our capacity to understand one another is diminished. Even when Maisel was working on this issue in the 1960s, he could see decreasing support for educational institutions as one consequence of the differentiated culture. That trend has continued.

It might take a different kind of journalism, backed by a different kind of financial support, to keep us together. For our social and political

2. This number is derived from circulation reported by the Newspaper Association of America in *Facts About Newspapers 2002* and household projections reported by the U.S. Census. Average daily circulation is counted by the formula $(6*D + S)/7$ where D = average weekday circulation and S = average Sunday circulation. For the historical perspective on household penetration, see Donald L. Shaw, "The Rise and Fall of Mass Media," Roy W. Howard Public Lecture, School of Journalism, Indiana University, April 4, 1991.

3. James Madison, Letter to W. T. Barry, August 4, 1822, in Saul K. Padover, ed., *The Complete Madison* (Millwood, N.Y.: Kraus Reprint Co., 1953).

4. John Merrill and Ralph Lowenstein, *Media, Message, and Man* (New York: David McKay, 1971).

health, we need to understand enough about the business of journalism to try to preserve it in new platforms.

The literature of business administration has a theoretical framework that provides a good place to start. Theodore Levitt popularized the term "disruptive technology" and captured the imagination of a generation of executives when he wrote "Marketing Myopia" for *Harvard Business Review* in 1960.[5] One of his examples came from the experience of the railroads. Their managers clung stubbornly to the narrow definition of their enterprise: they were in the railroad business. If only they had seen that they were in the transportation business, they might have been prepared when people and cargo began moving through metal tubes in the sky and along asphalt ribbons on the ground.

This model invites some rethinking of what business newspapers are, or should be, in. If you believe the Wall Street analysts most widely quoted in the trade press, newspapers are in the business of delivering eyeballs to advertisers. Everything not directly related to that is unrecovered cost.

Frank Hawkins was the director of corporate relations for Knight Ridder in 1986, a year when that group won seven Pulitzer Prizes. On the day of the announcement, the value of the company's shares fell. Hawkins called one of the analysts who followed the company and asked him why.

"Because," he was told, "you win too many Pulitzer Prizes." The money spent on those projects, the analyst said, should be left to fall to the bottom line.[6]

Knight Ridder at that time had another view. It was articulated for me in 1978 when I was posted from the Washington Bureau to corporate headquarters in Miami to serve as the company's first director of news research and help create an experimental electronic home information service. Hal Jurgensmeyer (1931–1995), a business-side vice president of the company, briefed me on the assignment. We were, he said, not in the news business, not even in the information business. We were "in the influence business." The sketch in Figure 1–1 is his.

5. Republished in Edward C. Bursk and John F. Chapman, eds., *Modern Marketing Strategy* (Cambridge, Mass.: Harvard University Press, 1964). Levitt's retrospective commentary appeared in *Harvard Business Review* 53 (September–October 1975), available at http://bold.coba.unr.edu/769/Week6/MarketingMyopia.pdf (retrieved June 29, 2003).

6. Personal conversation, Chapel Hill, N.C., March 22, 2002.

A newspaper, in the Jurgensmeyer model, produces two kinds of influence: societal influence, which is not for sale, and commercial influence, or influence on the consumer's decision to buy, which is for sale. The beauty of this model is that it provides economic justification for excellence in journalism.

Figure 1-1: The Influence Model

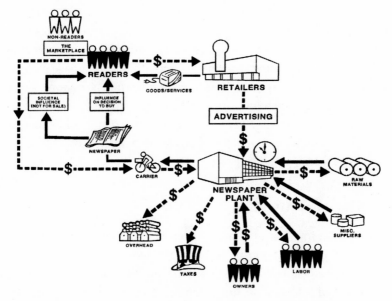

Hal Jurgensmeyer, 1978. Reproduced by permission of Knight Ridder.

A news medium's societal influence can enhance its commercial influence. If the model works, an influential newspaper will have readers who trust it, and therefore it will be worth more to advertisers.

Consider the supermarket tabloids. A front page that pretends to depict a presidential candidate chatting with an alien from outer space is going to attract only that limited subset of advertisers that depends on the most naively credulous subset of the population. A glance at the ads in such a newspaper bears out this supposition: ads for psychics and fake medical products such as pills that cause you to lose weight while you sleep.[7]

7. We all lose weight while we sleep, through evaporation, respiration, and elimination.

The disruption from technology in the newspaper case is more complicated than the railroad example and others used by Levitt. In those cases, the problem was one of straightforward product substitution. Cars were faster and more durable than horses, planes were faster than trains, natural gas was cheaper to process and transport than oil. For media, new technologies do all that, but they also change the nature of the audience. The new problem is overload on the ability of the audience to receive and consider the messages.

This phenomenon was under way long before the Internet appeared. Offset printing, which made it possible to create printing plates with a photographic process instead of hot lead, reduced the high fixed costs of publishing. Then computers made it easy to lay out a page at an editor's desk instead of by cutting and pasting in the back shop. The advances in printing technology opened the door to specialized publications with smaller audiences. Cheaper, slicker printing also made direct-mail advertising attractive and contributed to the demassification of the media long before there was an Internet.

The Scarcity of Attention

Herbert A. Simon saw the media overload problem coming. In one of a series of lectures sponsored by the Johns Hopkins Institute for Advanced Studies in Washington, D.C., in 1969–1970, he compared the surplus of information to his family's surplus of pet rabbits. They made the lettuce supply scarce. So it is with the wealth of information, which "means a dearth of something else: a scarcity of whatever it is that information consumes. What information consumes is rather obvious: it consumes the attention of its recipients.

"Hence a wealth of information creates a poverty of attention and a need to allocate that attention efficiently among the overabundance of information sources that might consume it."[8]

The national newspaper *USA TODAY* was one response to the problem. As originally designed, it let users scan the newspaper quickly and evaluate a large number of brief items. In most cases, the brief report

8. Herbert A. Simon, "Designing Organizations for an Information-Rich World," in Martin Greenburger, ed., *Computers, Communications, and the Public Interest* (Baltimore: Johns Hopkins University Press, 1971).

was enough. A reader might feel the need to follow up with other sources, but the newspaper had performed the service of alerting him or her. Harold Lasswell called this "the surveillance function." We all need a brief and efficiently presented heads-up about the dangers and opportunities that each new day presents.[9]

But few newspapers have copied that aspect of *USA TODAY*, and it has itself moved on to a more conventional mix of long and short stories. While there have been innovations in graphics and design for better information retrieval, the main response of the newspaper industry to the threat of substitute technology has been to reduce costs and raise prices. Some of the savings were achieved with better production methods. Others came from making the editorial product cheaper. Even as readership lagged, newspaper publishers congratulated themselves in the trade press for their ability to put out cheaper products while raising ad rates and subscription prices.

The strategy was possible because of newspapers' historically strong market position. Many newspaper owners—not all, of course—gained monopoly or near-monopoly situations by taking care to maintain the respect of their communities. For most of the twentieth century, newspapers were family businesses, run for the long term, attending to market share more than profitability. As the families sold to bigger organizations, the economics of publishing weeded out the weaker operations and built the eventual monopolies. Market share was no longer an issue. Yet for many this sense of obligation to community persisted.

The Harvesting Strategy

In recent years, as ownership has shifted from individuals and families to shareholders guided by professional money managers and market analysts, a shorter time horizon has come to dominate American business in general. The newspaper industry is no exception. At the same time, the newest of the disruptive technologies, online information service, may offer the most dangerous product substitution yet, especially in classified advertising. There is a textbook solution for a mature

9. Harold D. Lasswell, "The Structure and Function of Communication in Society," in L. Bryon, ed., *The Communication of Ideas: A Series of Addresses* (New York: Harper, 1948), 37–51.

industry that is unable to defend against a substitute technology. Harvard professor Michael E. Porter calls it "harvesting market position."[10] A stagnant industry's market position is harvested by raising prices and lowering quality, trusting that customers will continue to be attracted by the brand name rather than the substance for which the brand once stood. Eventually, of course, they will wake up. But as the harvest metaphor implies, this is a nonrenewable, take-the-money-and-run strategy. Once harvested, the market position is gone.

At the start of the twenty-first century, newspaper managers appeared to be harvesting—not so much for Porter's end-game strategy as to gather the financial resources to transfer the brand name to new ways of delivering news and advertising. Such a transfer requires investment, of course, and the problem was to generate the resources, without damaging the brand, in order to engage in costly experiments with new forms of media. But risk taking comes hard to an industry with such a long history of success, and the most interesting experiments with new media tended to come from elsewhere. And paying for new media ventures with shrinking news space and diminished reporting was itself a risky action, putting the newspaper industry's precious community influence in peril.

While today's investors might think it perverse, the notion of service to society as a function of business is neither new nor confined to those protected by the First Amendment. Henry Ford argued that profit was just a by-product of the service to society that his company performed. In 1916, Ford was sued by the Dodge brothers, Horace and John, who were counting on dividends from their 10 percent stake in Ford to invest in their own company. To their dismay, Ford had increased wages and cut prices at the same time. The brothers said he wasn't doing his duty to shareholders. Their counsel tried to trap Ford into admitting that.

"Your controlling feature, then," he asked Ford at the trial, "is to employ a great army of men at high wages, to reduce the selling price of your car so that a lot of people can buy it at a cheap price and give everybody a car that wants one?"

10. Michael E. Porter, *Competitive Strategy: Creating and Sustaining Superior Performance* (New York: Free Press, 1998), 311.

"If you give all that," said Ford, "the money will fall into your hands. You can't get out of it."[11]

The compatibility of profit and social virtue is not novel in capitalist theory. Such an absolutist as Milton Friedman has argued for "justly obtained profit."[12] And when newspapers were mostly owned by private individuals and families, the best of them tended to treat profitability as Ford did: as incidental to the main focus of business, which was making life better for themselves, their customers, and their employees.

Because a newspaper is so central to the functioning of its community, both for the commercial messages and the societal influence, the social pressure on a resident owner can be immense. And when publishers expanded into other markets, as the Knight brothers did from the base their father established in Akron, they tended to be careful about their standing in their new communities. Like Henry Ford, they had all the money they needed to meet their personal needs and desires. Ford sounded a lot like a newspaper publisher when he said all he wanted was "to have a little fun and do the most good for the most people and the stockholders."[13]

Ford was also assuring himself market share, which in a new technology business can be far more important than immediate profitability. In the early years, Ford Motor Company was only one of some 150 car makers. Like the dot-com entrepreneurs of today, he realized that to rise from the pack, he would have to create a product for a mass market.

When newspaper companies began going public in the 1960s, Wall Street was not immediately impressed. It was a mature industry, and the disrupting technologies were already apparent. Television had eaten into newspapers' share of national advertising, and cheaper, slicker printing was making it possible for highly specialized publications to thrive with narrowly directed advertising.[14] Moreover, the people who analyzed media businesses lived mostly in the large cities, where they

11. Carol Gelderman, *Henry Ford: The Wayward Capitalist* (New York: Dial Press, 1981), 84. Ironically, Ford Motor Company was far less successful when Ford was sole owner. After Ford bought out the minority shareholders in 1919, he began losing market share to General Motors and Chrysler.

12. For an excellent development of this idea, see John M. Hood, *The Heroic Enterprise: Business and the Common Good* (New York: Free Press, 1996).

13. Gelderman, *Henry Ford,* 85.

14. Maisel, "The Decline of Mass Media," 159.

could see newspapers dying or consolidating steadily since World War II. It took Al Neuharth of the Gannett Company to show them the profit potential of monopoly newspapers in small- and medium-size markets. These monopolies had not, for the most part, been managed to maximize near-term profitability. Their owners, partly out of a sense of social responsibility and partly with an eye on the long-term health of their companies, were more interested in influence than in maximizing their fortunes.

The Cowles family of Minneapolis and Des Moines is an example. When it sold *The Des Moines Register* in 1985 to Gannett Company, the paper covered the entire state of Iowa and had a tidy 10 percent operating margin. Gannett's finance people looked at the operation, saw no economic value in its statewide influence, and cut circulation back to the area served by advertisers in the Des Moines market. That saved money on the main variable costs, newsprint and ink. Two of five state news bureaus were eliminated. The operating margin went quickly to 25 percent.

Charles Edwards was publisher during the transition. "All these things, in and of themselves, one individual move wouldn't necessarily be devastating to the quality or the capacity of the newspaper to do good journalism," he said. "But collectively it had a huge impact . . . over time we just no longer had the capacity and the resources to do the kind of work we'd done."[15]

That ability to do journalism beyond the necessary minimum to maintain a platform for advertising, the capacity to do journalism with a broader purpose of maximizing a newspaper's influence—call it the influence increment of the business—existed in many markets. By systematically siphoning it off, the owners in these markets could give Wall Street the illusion that their mature and fading business was a growth industry. (Wall Street's myopic preoccupation with quarter-to-quarter changes in earnings helped, of course.)

W. Davis Merritt, former editor of *The Wichita Eagle*, tells a similar story. In the mid 1990s, Knight Ridder told him and his publisher to increase the operating margin to 23.5 percent.

"We looked and looked, and the only way we could do that was to

15. Charles Edwards, panel discussion, "Are the Demands of Wall Street Trumping the Needs of Main Street?" AEJMC Media Management and Economics Division, Miami Beach, Fla., August 8, 2002.

cut 10,000 circulation, reducing all our commitment anywhere west of Wichita. We had to tell 10,000 people who were buying and reading our paper, 'we're not going to let you buy our paper anymore.'"

That was damaging to Wichita, Merritt said, because the city's political influence in the state was a reflection of the newspaper's reach beyond its immediate market.[16]

There is a tendency among editorial-side people to blame these developments on the conversion of privately held companies to public trading. But more is going on than that. In 1984, Stanley Wearden and I compared the attitudes of newspaper managers (editors and publishers) with those of investment analysts toward the relative importance of financial performance and journalistic quality. We expected managers in public newspaper companies to agree with the analysts more than would the managers of nonpublic companies. We were wrong. At that time, managers in public and private companies held identical attitudes.

"King Kong with a Quotron"

A more important development, noted by Harvard's Rakesh Khurana, has been the gradual dispersion of ownership in corporate America in general—not just among media companies—from family and friends of the founders to institutional investors. In 1950, Khurana reported, less than 10 percent of corporate equities in the USA were owned by institutions such as pension funds and mutual funds. By the turn of the century, institutions controlled about 60 percent.

Some institutional investors, like Warren Buffett of Omaha, are long-term oriented. They call themselves "value investors" and think in terms of five years or even longer. But many, in the words of law professor Lawrence E. Mitchell, are like "King Kong with a Quotron." The typical money manager "is paid on the basis of how much money he makes for the fund in the short run, a fact which focuses his attention on the short-term performance of his portfolio corporations."[17] Like the Dodge

16. W. Davis Merritt, remarks to Seminar in Media Analysis, School of Journalism and Mass Communication, University of North Carolina at Chapel Hill, Chapel Hill, N.C., April 22, 2002.

17. Lawrence E. Mitchell, *Corporate Irresponsibility: America's Newest Export* (New Haven, Conn.: Yale University Press, 2001), 170. An analysis by Rick Edmonds indicated that newspaper companies had a disproportionate number of value investors. See

brothers attacking Henry Ford, these kinds of institutional investors wanted faster payout from their holdings, and they were getting aggressive about it by the 1980s. They promoted hostile takeovers and pressured directors to be more responsive to short-term interests of shareholders.[18]

An even broader view is taken by Jane Cote of Washington State University–Vancouver, who sees investor pressure eroding professionalism in a variety of fields regardless of their business structure.[19] Doctors who sell their practices to corporations risk losing control of their ability to maintain professional responsibility, whether the corporations are public or private. The corruption of the professional values of accounting firms, made highly visible by Enron and related scandals in 2002, showed that a company didn't have to be a corporation at all to be affected. Accounting firms are partnerships, not corporations. The erosion of professional values might be a useful frame for examining what is happening to newspaper journalism.

But the most interesting frame is the influence model. If the influence model is valid, then newspaper companies that yield to investor pressure to convert the influence increment into cash have either decided to harvest their market position and get out, or they are taking a terrible risk. They are trying to harvest enough of their market position to provide the capital to move into new forms of media without making any new investment. But new enterprises do require investment. The best way to ensure the future of newspapers would be to conserve their influence and pay the costs of the radical experimentation needed to learn what new media forms will be viable.

These still developing media forms are the real competitors, and market share is an issue again. It is a more complicated market because the good being sought is neither share of circulation nor share of readership, but rather share of the audience's finite amount of attention. Like Henry Ford, media entrepreneurs, including newspaper companies,

---

"Who Owns Public Newspaper Companies and What Do They Want?" Poynter Online, www.poynter.org, December 11, 2002 (retrieved December 10, 2003).

18. Khurana Rakesh, *Searching for a Corporate Savior: The Irrational Quest for Charismatic CEOs* (Princeton, N.J.: Princeton University Press, 2002).

19. Seminar in Media Analysis, University of North Carolina, Chapel Hill, N.C., March 20, 2002.

should be more interested in capturing the relevant share than in max-imizing short-term profitability.

How can we test the reality of the influence model? A crude first step would be to find a measure of something that relates to newspaper in-fluence to see if it changes over time along with readership. The Gen-eral Social Survey offers trend data on both readership and confidence in the press. First, let's look at confidence:

Figure 1-2

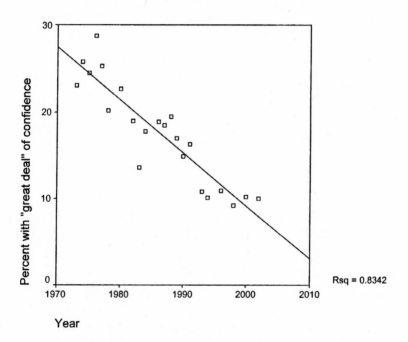

Confidence in the Press 1972-2002

The trend line falls at an average rate of 0.6 of a percentage point per year, which would take it to zero in 2015. But as you can see, the de-cline shows signs of leveling off after a sharp drop between 1991 and 1993. Now let's see what has happened to the daily newspaper reading habit in that same period.

This is a steeper line, and there is less year-to-year variation around it. The slope is a bit more than 0.95 percentage points per year. Try

Figure 1-3

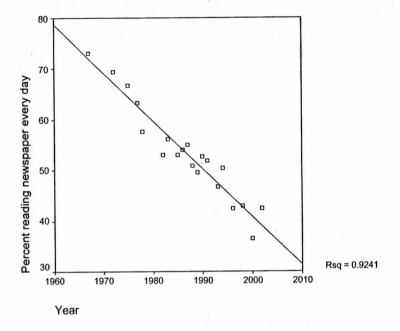

Daily newspaper readership 1967-2002

Rsq = 0.9241

extending that line with a straightedge, and it shows us running out of daily readers late in the first quarter of 2043.[20]

The fact that both confidence and readership are declining in the same period and at close to the same rate does not mean that one is the cause of the other. Time is one of those variables that Harvard Professor James A. Davis calls "fertile" because they produce lots of different things. When we take out the effect of time (with a partial correlation of readership and confidence, controlling for year) we are left with a very low and not at all significant correlation.[21]

The true source of the readership decline can easily be seen in the

20. This plot has a data point for 1967 when NORC asked the newspaper readership question for the Nie-Verba "Participation in America" survey and 73 percent reported reading a newspaper every day. The General Social Survey was started in 1972.

21. James A. Davis, *The Logic of Causal Order* (Newbury Park, Calif.: Sage Publications, 1985). The correlation results are $r = .164$, $p = .544$.

next graph. For years, marketers have been acting as though it is a big mystery. Instead, the basic problem is quite simple. It is a matter of generational replacement. Since the baby boomers came of age, we have known that younger people read newspapers less than older people. For years, we comforted ourselves with the thought that they would become like us and adopt the newspaper habit as they got older.[22] It never happened. In Figure 1–4, each line is a generation. The lines are roughly parallel (allowing for some kinks due to sampling error). As the years go by, each generation keeps roughly the reading habit that it had established by the time it reached voting age. (Although you can see the aging World War II generation start to falter in 2002. Its youngest members were then 74.)

Figure 1-4

Generational Change in Newspaper Readers

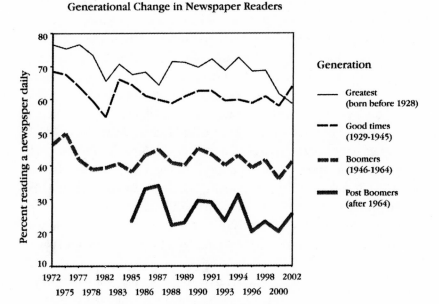

22. For example, Heidi Dawley of *Media Life Magazine,* quoting Dr. Leo Kivijarav, then director of research and publication for Veronis Suhler Stevenson, the investment banking firm: "Kivijarv believes that as people get older they take on the habits of older people. So as the young folks today who spurn newspapers get older, they will become newspaper readers." "Dispelling myths about newspaper declines," *Media Life Maga-*

Let your gaze linger on the above chart so the message sinks in. Here
it is in straight words: all the readership studies to learn "what readers
want" and all the resulting tweaking of content in response, all that activity throughout the final quarter of the twentieth century, didn't matter. The important things that affect readership are happening before
the customers are old enough to show up in reader surveys!

Well, almost. We're talking about daily readership here, and you might
still make a pretty good business out of a product that readers consult
only two or three times a week. Let's go back to the influence model
and try to determine if it is viable in such a world.

The trouble with most survey analysis is that it is based on individual
responses. But what we really care about is the public sphere and the
newspaper's role in creating it. There might be some happy combination of individual and community attributes that makes the influence
model work. To find out, we need an experimental design that compares newspaper use in communities with different levels of credibility
over an extended period of time. No such data set exists. However,
historical data are available for newspaper circulation and household
penetration, and the Knight Foundation has begun a time series in
twenty-six communities—with measures every three years—that includes a question on newspaper believability. If we can accept credibility,
measured by a single survey question on believability, as an indicator of
newspaper influence, then we can at least make a start.

Editors have been worrying about the credibility issue for years. The
most alarming report on the topic came in 1985 from Kristin McGrath
of MORI Research with a little nudging by David Lawrence who, on behalf of the American Society of Newspaper Editors, hired her firm to do
a national survey. "Three-fourths of all adults have some problem with
the credibility of the media," she wrote, "and they question newspapers
just as much as they question television."[23]

A contrasting report was issued early the following year by the Times
Mirror Company after it hired the Gallup Organization to cover the
same territory. "If credibility means believability, there is no credibility
crisis," said this report, written by Andrew Kohut and Michael Robin-

azine, June 4, 2003. www.medialifemagazine.com/news/2003 (retrieved December 10,
2003).

23. *Newspaper Credibility: Building Reader Trust, a National Study Commissioned by the
American Society of Newspaper Editors* (Minneapolis: MORI Research, Inc., April 1985).

son. "The vast majority of the citizenry thinks the major news organizations are believable."[24]

Oddly, the data collected by the two organizations were not very different. Their different interpretations reflected more a half-full versus half-empty attitude difference than a data contrast.[25]

Another contribution to the conversation came in 1998 when Christine Urban, also working for ASNE, produced another report. Hers made no reference to the earlier work, but it did propose six major sources of low trust. Number one on the list: "The public and the press agree that there are too many factual errors and spelling or grammatical mistakes in newspapers."[26]

Two purely descriptive studies were published in 2001. News credibility was one of a very broad array of social indicators measured in a 1999 Knight Foundation survey which found that 67 percent believe "almost all or most" of what their local daily newspaper tells them. A similar result was published at the same time by *American Journalism Review*, based on field work in 2000 funded by the Ford Foundation. This study reported that 65 percent believe all or most of what they read in the local paper.[27]

Designers of none of these studies made any effort to attain compatibility with previous work so that comparisons could be made over time. Nor were any of the studies informed by any kind of theory that might help us understand how much credibility a newspaper needs, how much it costs to get it, and whether the cost is worth it. As careful and detailed as they were, these reports generated little but description "waiting for a theory or a fire."[28]

24. *The People and the Press: A Times Mirror Investigation of Public Attitudes toward the News Media Conducted by the Gallup Organization* (Washington, D.C.: Times Mirror, January 1986).

25. Philip Meyer, "There's Encouraging News About Newspapers' Credibility, and It's in a Surprising Location," *presstime*, June 1985.

26. Christine D. Urban, *Examining Our Credibility: Perspectives of the Public and the Press* (American Society of Newspaper Editors, 1989), 5.

27. *Listening and Learning: Community Indicator Profiles of the Knight Foundation Communities and the Nation* (Miami: Knight Foundation, 2001); Carl Sessions Stepp, "Positive Reviews," *American Journalism Review*, March 2001.

28. This phenomenon is not confined to media businesses. Ronald Coase, in a critique of early institutional studies in business administration, said, "Without a theory they had nothing to pass on except a mass of descriptive material waiting for a theory or a fire." Quoted by Oliver E. Williamson in Giovanni Dosi, David J. Teece, and Josef

The appeal of the influence model is that it provides a business rationale for social responsibility. The way to achieve societal influence is to obtain public trust by becoming a reliable and high-quality information provider, which frequently involves investments of resources in news production and editorial output. The resulting higher quality earns more public trust in the newspaper, and not only larger readership and circulation but also influence with which advertisers will want their names associated.

Because trust is a scarce good, it could be a natural monopoly. Once a consumer finds a trusted supplier, there is an incentive to stay with that supplier rather than pay the cost in time and effort to evaluate a substitute. I've been going to the same barbershop for twenty years for that reason.

It follows, then, that societal influence of a newspaper achieved from practicing quality journalism could be a prerequisite for financial success. Over the long term, social responsibility in the democratic system supports, rather than impedes, the fulfillment of a newspaper's business objectives, through the channels of obtaining public trust and achieving societal influence, which then feeds back into further fulfillment of the public mission, thereby creating a virtuous cycle (see Figure 1–5).[29]

Reversing the argument, cutbacks in content quality will in time erode public trust, weaken societal influence, and eventually destabilize circulation and advertising. So why would anyone want to cut quality? If management's policy is to deliberately harvest a company's market position, it makes sense. And pressure from owners and investors might even lead managers to do it without thinking very much about it because reducing quality has a quick effect on revenue that is instantly visible, while the costs of lost quality are distant and uncertain. When management is myopically focused on short-term performance indicators such as quarterly or annual earnings, the harvesting strategy is understandable.

If those distant costs could be made more concrete and predictable, managers and investors might make different decisions. This is a nat-

---

Chytryl, eds., *Technology, Organization and Competitiveness: Perspectives on Industrial and Corporate Change* (New York: Oxford University Press, 1998).

29. This is the model proposed in Philip Meyer and Yuan Zhang, "Anatomy of a Death Spiral: Newspapers and Their Credibility," AEJMC, Miami Beach, Fla., August 10, 2002.

Figure 1-5: Societal Influence Model for the Newspaper Industry

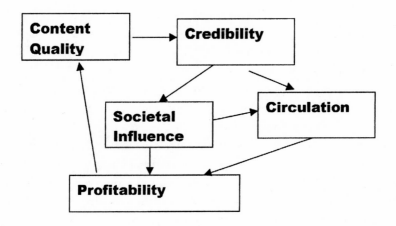

ural agenda for journalism researchers: to reduce the uncertainty about the long-term cost of low credibility, using individual communities and their newspapers as the level of analysis.

Testing the Model

The first test of the model is very simple: find a correlation between credibility and profitability. After that, it gets difficult. Correlation does not prove causation, so we'll have to treat this as a puzzle and include a dose of common sense in the solution. For starters, we need to be able to measure both credibility and profitability at the level of individual newspapers. Fortunately, a convenience sample is available.

The Knight Foundation keeps track of the twenty-six communities where John S. and James L. Knight operated newspapers in their lifetimes.[30] (This list of twenty-six only partially overlaps with newspapers owned by Knight Ridder. While the company buys and sells newspapers, the foundation's list remains constant.) Knight Ridder's communities range from large (Philadelphia and Detroit) to very small (Milledgeville, Georgia, and Boca Raton, Florida). One of the questions in its 1999 and

30. The John S. and James L. Knight Foundation promotes excellence in journalism worldwide and invests in the vitality of twenty-six U.S. communities where the communications company founded by the Knight brothers published newspapers. The foundation is wholly separate from and independent of those newspapers.

2002 surveys asks this question about the communities' general trust in newspapers as a media category: "Please rate how much you think you can believe each of the following news organizations I describe. First, the local daily newspaper you are most familiar with: would you say you believe almost all of what it says, most of what it says, only some, or almost nothing of what it says."

It is a crude measure compared to Gaziano and McGrath's multi-item measure for ASNE in 1985. Its value comes from its consistent use in multiple markets at two points in time.[31]

Profit data is hard to get, so we'll use a special measure of circulation as a stand-in. One advantage to doing so is that circulation is fairly stable, while profit depends on fluctuations in the business cycle. To adjust for different market sizes, let's use circulation divided by the number of households. Newspaper marketers call this number "market penetration." That leaves yet another problem, how to define the market. Newspapers vary wildly in their self-definition of markets. Sometimes, their definitions change, as in the Des Moines case. While maximizing circulation is not always the business goal, I know of no important newspaper that is ambivalent about circulation in its home county. Therefore, it makes sense to take home county penetration as our uniform indicator of business success.

One more step is needed to put matters on an apples-to-apples basis. Some markets have chronically low penetration due to local circumstances. We can adjust for such built-in differences by making change over time the key indicator. Each market defines its own baseline, and business success is measured by the ability to improve on that base—or minimize the decline—from some reference point in time. The result is a fair comparison of audience response to changing circumstances across wildly different markets.

Such a definition deserves a name. I call it "penetration robustness," and I calculate it from the 1995, 2000, and 2003 county penetration reports of the Audit Bureau of Circulations (ABC). Penetration declined almost everywhere. But it makes sense to call it robust when its value at Time 2 is a relatively high proportion of its value at Time 1.[32]

31. *Listening and Learning.*

32. Such a measure is complicated by the variance in the lag time between an ABC audit and publication of the result in the reports. Each report presents results of the

So much for theory. Now to some practical obstacles. Use of ABC's county penetration report made it necessary to eliminate two Knight communities where the 1999 survey geography was not defined by counties.[33] A third, Milledgeville, had to go because it was not an ABC client.

That leaves a sample of twenty-three markets. In south Florida, the sample size was great enough to allow separation of Dade and Broward counties so that they could be treated as separate communities. Now we have twenty-four markets, most of them with dominant newspapers that are now, or have been, owned by Knight Ridder. Palm Beach and Broward counties in Florida are exceptions. In the former, the dominant paper is Cox's *Palm Beach Post*. In the latter, it is the Tribune Company's *South Florida Sun-Sentinel*. Several counties have more than one strong newspaper. Because we are more interested in a test of influence theory than in the fate of individual newspapers, having multiple cases in a county can actually help reduce error. We'll use county-level aggregates for both robustness and credibility in the hunt for a correlation.

Here's another practical obstacle: the single-question credibility measure in the Knight Foundation survey is unstable. There was considerable shifting among the counties. The correlation between the 1999 and 2002 measures was significant but lower than we would like. Sedgwick County, Kansas, gained seven points while Grand Forks, North Dakota, lost six. Transient local controversies or major news events might have had something to do with these shifts.[34]

Accordingly, it seems prudent not to base the credibility measure on a single reading but to average the 1999 and 2002 survey findings. This yields a range from 15 percent trusting their "most familiar" newspaper

---

most recent audit, which can be anywhere from a few months to more than a year before the date of the report. To correct for this, I used the number of days elapsed between audits to calculate an annualized robustness. It can be conceptualized as the proportion of home county penetration retained in an average year during the study period. To summarize: Penetration = circulation/households. Robustness = Penetration at Time 2/Pentration at Time 1. Annualized Robustness = $1-(1-R) / A$ where R = Robustness for the period and A = elapsed time in years between ABC audits.

33. Long Beach, California, and Gary, Indiana.

34. The correlation numbers were $r = .457$, $p = .022$. Despite individual shifts, the twenty-five Knight counties were stable as a group. The percent who believed their newspaper all or most of the time averaged 20.4 percent in 1999 and 20.7 percent in 2002.

among the six circulating in Boulder County, Colorado, to 27 percent among readers of the three dailies sold in Brown County, South Dakota. It also yields a credibility distribution with no outliers or extreme cases.[35]

Alas, the same cannot be said for the circulation data. There were three outliers in the 1995–2000 period.[36] Examining each of them in turn, I found local factors affecting circulation that would overwhelm any more subtle effects that we might seek:

• Dade County, Florida. The county's explosive circulation boom was the result of an artifact, the unbundling of *El Nuevo Herald* from its mother ship, *The Miami Herald*. Before the separation, ABC folded circulation of the Spanish language edition into the *Herald's* total. After the split, *Herald* circulation went down, but counting *El Nuevo* separately made total circulation rise sharply. There was no way to correct for this to make a before-after comparison, and Dade County was dropped from the sample.[37]

• Boulder County, Colorado. In the months before the creation of the joint agency by the owners of *The Denver Post* and the *Rocky Mountain News* in 2000, the two Denver newspapers were engaged in a bitter circulation war that saw the price of a newspaper drop to a penny per copy. This battle extended into neighboring Boulder County. While it cost the local paper circulation, total newspaper circulation in the county soared. Boulder County was dropped.[38]

• Wayne County, Michigan. Detroit, always a strong labor town, underwent a bitter newspaper strike that began in 1995 and led to many

35. Using the Tukey boxplot, the outliers were Brown County in 2002 and Leon County, Florida, and Grand Forks, North Dakota, in 1999. The six papers circulating in Boulder County in 2000 were the *Daily Camera, Denver Post, Denver Rocky Mountain News, Fort Collins Coloradan, Daily Times-Call,* and *USA TODAY.* In Brown County the three were the *Aberdeen American-News, USA TODAY,* and the *Star Tribune (Newspaper of the Twin Cities).*

36. All four met Tukey's definition of outliers and extreme values as cases that are more than 1.5 times the interquartile range from the upper and lower edges of that range. See John W. Tukey, *Exploratory Data Analysis.*

37. I appreciate the help of Armando Boniche, research manager, *The Miami Herald,* in sharing this history.

38. Barrie Hartman, former executive editor of the *Boulder Daily Camera,* provided this background.

union members losing their jobs. In a display of sympathy and solidarity, enough working people in the home county stopped buying the paper to cause a catastrophic circulation decline. Wayne County was removed from the sample. That leaves twenty-one communities without obvious unusual circumstances to mask the effect of credibility on circulation.

Just for reassurance that we are grounded in the real world, here is the list of twenty-one counties with their credibility scores and average annual robustness for the period 1995–2000. Please remember that these are aggregate scores for all the newspapers in the county. Credibility comes from the survey question about the newspaper read most often, and robustness comes from ABC data for all audited newspapers in the county. They are listed in ascending order of their credibility.

Table 1-1: County credibility and robustness

County	Credibility	Robustness
Leon FL	16.23	.9561
Philadelphia PA	17.00	.9860
Summit OH	17.26	.9798
Mecklenburg NC	18.27	.9817
Ramsey MN	18.39	.9762
Allen IN	19.47	.9797
Bibb GA	19.82	.9820
Palm Beach FL	19.91	.9824
Manatee FL	20.14	.9793
Sedgwick KS	20.20	.9687
Santa Clara CA	20.69	.9772
Richland* SC	20.78	.9647
Centre PA	20.85	.9844
Fayette KY	21.43	.9885
St. Louis MN	21.76	.9760
Horry SN	22.12	.9838
Broward FL	22.43	.9749

Table 1-1: County credibility and robustness (*cont.*)

County	Credibility	Robustness
Muscogee GA	24.32	.9930
Harrison MS	24.33	.9814
Grand Forks ND	26.46	.9904
Brown SD	27.15	.9931

*Includes Lexington County in 1999 survey

The plot in Figure 1–6 shows that credibility and penetration robustness are strongly related. As one rises, so does the other. Their relationship is quite unlikely to be due to chance. The credibility of the newspapers in these communities explains 31 percent of the variation in the robustness of their combined daily penetration. In other words, if you tell me the credibility of the newspapers in any given county, I can estimate their combined robustness with 31 percent greater accuracy than if I just used the twenty-one-county average to inform my guess.

Each little square in the plot is a county, and I have identified some as space permits.[39] As credibility increases from left to right, robustness rises from bottom to top.

The slope of a straight line defining that relationship is 0.2, meaning that annual circulation robustness—the ability of a county's newspapers to hold their collective circulation in the face of all of the pressures trying to degrade it—increases on average by two-tenths of a percentage point for each 1 percent increase in credibility. However, there is a problem. Just by inspection of the list, one can see that credibility tends to be greater in the smaller markets. Keith Stamm, the University of Washington professor who investigated the effect of community ties on newspaper use, prepared us for the possibility that those ties might weaken as cities grow.[40] Distrust of local newspapers could be a function of weaker community ties.

The size effect is not linear because it diminishes as counties get

39. R = .555, p = .009.
40. Keith Stamm, *Newspaper Use and Community Ties: Toward a Dynamic Theory* (Norwood, N.J.: Ablex, 1985). He reports contrasting arguments at pp. 178–79.

Figure 1-6: Circulation robustness by credibility

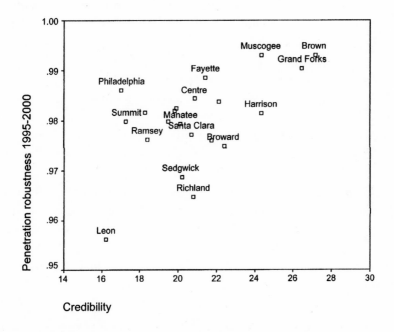

larger. For counties larger than four hundred thousand households, there is no size effect at all. One way to deal with a curvilinear effect like this one is to reexpress market size as its logarithm. The correlation between the log of households (which takes the curve into account) and credibility is negative, significant, and powerful.[41] An effect this basic needs to be noted so that we can take it into account when we look at other aspects of newspaper quality. We should start to suspect that large and small markets are qualitatively different from one another.

Looking for Causes

Now for the hard part, showing the path of causation behind these interesting connections. We have three nicely linked variables: market size, credibility, and robustness. Which are causes and which are effects? While this might look like a statistical problem, it's really more of a logic issue. The most important logical rule is that tomorrow's events

41. For log households and credibility, $r = -.609$, $p = .003$.

can't cause today's.[42] A related principle is that events closest to one another in a causal sequence will have the highest correlations. The other common-sense rule is to identify what Jim Davis calls "fertile variables" and "sticky variables." The sticky variables are those that do not change easily, if they change at all. One's year of birth is an example. Such variables are more likely to be causes than effects. Fertile variables are those that affect everything they touch—social class, for example. This characteristic makes us look to them first when seeking causes.

In this case, market size is both sticky—it does not change quickly—and fertile—it has lots of effects. So there is plenty of reason to think of it is as a first cause in this simple model. To help understand the situation, here's another picture. In Figure 1–7, the numbers show the correlation coefficients while the direction of the arrows gives our best guess at the direction of causation.

Figure 1-7: The intercorrelations of market size, quality, and robustness

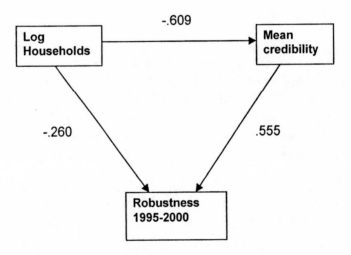

What this picture says is that market size causes credibility (the sign is negative, so smaller communities tend to be more trusting), and credibility causes robustness. There appears to be a smaller, direct effect of

42. Or, as Jim Davis has put it, "After cannot cause before." James A. Davis, *The Logic of Causal Order* (Beverly Hills, Calif.: Sage Publications, 1985), 11.

market size on robustness. Or is there? It might be that the small correlation of -.260 is just an artifact of the reduced credibility in larger markets.

There is a statistical way to help us decide. It is called partial correlation. It tells us what the correlation between market size and robustness
would be if all the markets had the same credibility. If you like sports
metaphors, you can call this "leveling the playing field." If you want to
sound like a statistician, then say you are controlling for credibility.

Figure 1-8: Path analysis with controls

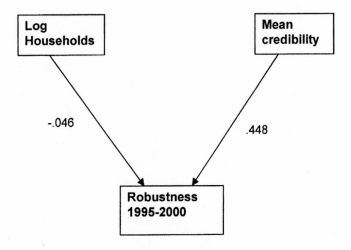

When market size (log of households) is controlled, the correlation
between credibility and robustness remains positive and significant.
That gives us confidence that credibility, in and of itself, improves circulation robustness. But when credibility is controlled, the link between size and robustness virtually disappears. This leads us to believe
that size itself doesn't matter except as it affects credibility. The path is
straightforward: smallness in size contributes to credibility, which in
turn aids robustness. The direct effect of size is nil.

This result is consistent with Maisel's thesis that media aimed at
smaller audiences do better, and it hints that credibility achieved by
narrower targeting is the reason. It is also very good news for editors.
They can't do much about the size of their home counties, but they can

at least try to imagine ways to manage a larger newspaper that would yield some of the effects of a smaller community. Zoning is one obvious way. Encouraging citizen participation in the affairs of the larger community, a goal of the civic journalism movement, is another.

What we have here is evidence in support of one link in the model in Figure 1–3. It is weak evidence, limited as it is to twenty-one markets and one five-year time period. When the data are used to predict robustness from 2000 to 2003, nothing happens. Perhaps it takes longer than three years for the effect of trust to accumulate to the point of being measurable. However, this is not true of all effects, as we'll see in chapter 5.

This case illustrates the difficulty of demonstrating, in a way that will be appealing to investors, the relationship between quality journalism and profitability. Hal Jurgensmeyer's influence model deserves further investigation, but several things are needed:

1. Bigger samples and longer time periods. Teasing out the direction and degree of causation requires looking at more points in time so that effects can be separated from ongoing secular trends. It also needs a more varied sample. Because of their history, the communities tracked by the Knight Foundation tend to have better than average newspapers. With more variance in newspaper quality, we might see stronger effects.

2. A more stable measure of influence. The 1985 ASNE study provided evidence that a newspaper's perceived ties to its community are a factor in its credibility, and by implication, its influence. That lead should be pursued. The concept is too important to rely on a single survey question.

3. Recognition that small and large papers operate in environments so different that they probably need to follow different rules. Other research has shown that circulation volatility is much greater among smaller newspapers.[43] If the rules are different, somebody needs to discover and codify them.

4. Identification of the "sweet spot" and a means of determining where a given newspaper stands in relation to it.

43. Philip Meyer and Minjeong Kim, "Above-Average Staff Size Helps Newspapers Retain Circulation," *Newspaper Research Journal* 24:3 (Summer 2003), 76–82.

The last one requires some explanation. The concept of the sweet spot has been used by Jack Fuller to define an optimal compromise between demands of profitability and public service. My definition is a little different. It assumes that quality brings in more dollar return than it costs—up to a point. It is vital for a manager—or an investor—to know whether that point is approached, reached, or exceeded. Imagine a bell-shaped curve with profit measured along its horizontal dimension. Here is a picture of such a curve.

Figure 1-9

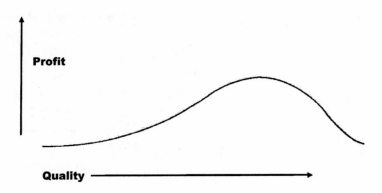

Quality is measured on the horizontal. As it increases, so does profit—slowly at first, then, as critical mass is reached, with accelerating effect. Then the curve rounds off at the top and begins to descend. This is the law of diminishing returns. Since it's a natural law, there's probably not much we can do about it. But we can try to determine when and how the diminution of return sets in.

This goal is much harder than it sounds because the effect of quality takes some time to kick in. The initial effect is cost. The monetary return comes later after the newspaper's influence accumulates. The good news is that there is a similar lag when quality is degraded. Readers and advertisers alike keep using the product out of habit long after the original reasons for doing so are forgotten.

A newspaper is in danger when its owners assume that it has passed the sweet spot and is on the downhill side of the curve. If they have really passed the point of diminishing return, they would be right to cut quality and get back to a level that makes economic sense. But what if they are in fact on the uphill side? In that case, the quality cuts will drag them backward down the slope to eventual destruction.

Since the dawn of the electronic era in the 1920s, newspapers have minimized their decline by adapting to a long string of new technologies that disrupted their then existing business models: radio, television, high-quality printing for direct-mail advertising and highly specialized print media. Of the six classic strategies enumerated by Michael Porter for dealing with competitive new technology, four clearly apply to newspapers:[44]

1. Enlist suppliers to help in defense. Newsprint suppliers were major funders of the newspaper industry's experiments with market research in the 1960s and 1970s. United Press International, a supplier of news, chipped in even when it was in serious financial trouble itself.[45]

2. Redirect strategy toward segments that are least vulnerable to substitution. When television captured national image advertising, newspapers concentrated more on detailed price and product advertising for local retailers. After direct-mail innovators put advertising messages on slick paper and in color, newspapers responded with preprints and total market coverage (TMC) products.

3. Enter the substitute industry. Before the FCC limited cross ownership, some newspaper companies were quick to acquire and apply their brand name to radio and TV stations. Now they are responding to the Internet threat by creating online versions of themselves.

4. Harvest market position. This is the "take-the-money-and-run" plan. Because newspaper customers are such creatures of habit, it could be quite seductive. It means raising prices, reducing quality, and taking as much money out of the firm as possible before it collapses.

44. The six possible strategies are listed by Porter in his chapter on substitution (chapter 8) in *Competitive Advantage*. Two lack an obvious application to newspapers. One of these is finding new uses unaffected by the substitute, such as the use of baking soda as a refrigerator deodorant, an application that now exceeds the product's use for baking. The other is to compete in areas where the substitute is weakest. For newspapers, applications requiring the portability of the printed product might have some promise.

45. These efforts are chronicled by Leo Bogart in *Preserving the Press: How Daily Newspapers Mobilized to Keep Their Readers* (New York: Columbia University Press, 1991).

I know of no newspaper company that is totally committed to the fourth strategy. But on some days, at some companies, there are very strong indications that they are drifting in that direction, egged on by short-sighted investors. Documentation and detailed specification of the Jurgensmeyer influence model might help the industry avoid this fate.

Or it might not. Even if nothing can save newspapers from the leaders that William Allen White feared, the ones with acquisitive faces, the influence model could motivate the creators of the newer forms of media to find a way to keep social responsibility in their business plans. It deserves our highest priority. We should have started sooner.

2

How Newspapers Make Money

I learn'd to love despair.
 —Byron, "The Prisoner of Chillon," 1816

T O A P P R E C I A T E how the locale for Byron's poem serves as a metaphor for the American newspaper industry, drive a short distance north of Montreaux, Switzerland, and visit the Rock of Chillon. It sits on the eastern edge of Lake Geneva, and it was fortified in the ninth century. In the twelfth century, the counts of Savoy built a castle on the rock. With the lake on the west side and a mountain on the east, the castle commands the north-south road between. Any traveler on that road had to choose between paying a toll to the owner of the castle, climbing the mountain, or swimming the lake. It was such a sweet deal that the lords of Savoy and their heirs clung to that rock for three centuries.

For generations, U.S. newspa-

An earlier version of this chapter appeared as "Learning to Love Lower Profits," *American Journalism Review* 17:10 (December 1995), 40–44.

34

per publishers were like the Savoy family. Their monopoly newspapers were tollgates through which information passed between the local retailers and their customers. For most of the twentieth century, that bottleneck was virtually absolute. Owning the newspaper was like having the power to levy a sales tax.

But new technology is bypassing the bottleneck. Just as today's travelers can fly over Chillon or bypass it in a power boat or on an alternate road, today's retailers are finding other ways to get their messages out. Newspapers have been slow to adapt, because their culture is the victim of that history of easy money. For perspective, consider the following comparison: in most lines of business there is a relationship between the size of the profit margin—the proportion of revenue that trickles to the bottom line—and the speed of product turnover. Sell a lot of items, and you can get by with just a little profit on each one. If sales are few, you'd better make a bundle every time.

Supermarkets can prosper with a margin of 1 to 2 percent because their buyers consume the products continually and have to keep coming back. Sellers of diamonds or yachts or luxury sedans build much higher margins into their prices to compensate for less frequent sales. Across the whole range of retail products, the average profit margin is in the neighborhood of 6 to 7 percent.

In turnover, daily newspapers are more like supermarkets than yacht dealers. Their product has a one-day shelf life. Consumers and advertisers alike have to pay for a new version every day if they want to stay current. Absent a monopoly, newspaper margins would be at the low end. But when they owned the bottleneck, the opposite was true. Before technology began to create alternate toll routes, a monopoly newspaper in a medium-size market could command a margin of 20 to 40 percent.

That easy-money culture has led to some bad habits. If the money comes in no matter what kind of product you turn out, you become production-oriented instead of customer-oriented. You are motivated to get it out the gate as cheaply as possible. If your market position is strong, you can cheapen the product and raise prices at the same time. Innovation happens, but it is often directed at making the product cheaper instead of making it better.

Before newspapers were controlled by publicly held companies, their finances were secret. A few retailers may have noticed that a publisher's family took flying vacations to Europe while they drove theirs to the local mountains or the beach, but publishers were usually careful not to flaunt their wealth. It is unwise to arouse resentment from one's own clientele.

When newspaper companies began going public in the late 1960s, the books opened, and Wall Street was delighted with what it saw. Analyst Patrick O'Donnell, in an industry review prepared for E. F. Hutton clients in 1982, ticked off the advantages enjoyed by newspapers:

• Competition in selling was increasing the size of the advertising market, and newspapers had consistently received 28 to 30 percent of total advertising dollars in the previous two decades.

• Newspapers were practically immune from the profit-eroding effects of inflation because they could "pass cost pressures along through prices very efficiently."

• With their strong cash flows, newspaper companies could finance growth with their own money and avoid the uncertainties of fluctuating interest rates.[1]

Looking at the previous decade, O'Donnell noted with satisfaction that "aggressive pricing met little resistance, especially in one-newspaper markets where retailers have few other means of access to their customers." (Note his use of the word "aggressive" to describe increasing prices. In other industries discussed in business literature, aggressive pricing is *lower* pricing. That's because most industries are competitive, and a price cut is aggression against the competition. Newspapers, being mostly monopolies, direct their price aggression against their customers instead of each other.)

At the same time, newspapers were using technology to bring their costs down. Production costs decreased in the last quarter of the twentieth century as the change to cold type and automated typesetting was completed. Circulation costs diminished when newspapers pulled back their "vanity circulation" in areas not considered important by the advertisers in the newspapers' retail trade zones. Administrative costs went up. And news-editorial costs became more flexible.

In 2001, a typical newspaper of one hundred thousand in circulation

1. Patrick O'Donnell, "The Business of Newspapers: An Essay for Investors," E. F. Hutton Equity Research, February 12, 1982.

would have an expense breakdown like this (dollar amounts are in millions: percents are of revenue):[2]

News-editorial	$5.36	11.4%
Advertising	4.00	9.4
Circulation	5.68	11.1
Production	3.99	8.0
Newsprint and ink	6.43	13.9
Administration and depreciation	12.43	25.4
Gross profit	10.98	20.8
Total	55.48	100

The revenue side was much simpler: Industry-wide, advertising accounted for 82 percent of newspaper revenue in 2000 and circulation was the other 18. That was a shift from a 71–29 division in midcentury. Within the three main categories of advertising—national, retail, and classified—the latter grew in relative importance. Here is media economist Robert Picard's analysis of the change between 1950 and 2000:[3]

	1950	*2000*
Retail	57%	44%
National	25	16
Classified	18	40

That, as Picard suggested, is a whole new business model. And it is a less stable model. Classified advertising, because it includes help-wanted, real estate, and automobile advertising, is especially subject to changes in the business cycle.

Wall Street sees the cyclical nature of the newspaper business as a major drawback. The financial analysts who advise institutional investors make their reputations on their ability to predict the future. So they prefer companies whose growth patterns are steady year in and year out.

2. These numbers are based on grouped data compiled by the Inland Daily Press Association, "National Cost and Revenue Study for Daily Newspapers for 2001."

3. Robert G. Picard, *Evolution of Revenue Streams and the Business Model of Newspapers: The U.S. Industry between 1950–2000* (Business Research and Development Center, Turku School of Economics and Business Administration, Turku, Finland, 2002), 25. Also published in *Newspaper Research Journal* 23:4 (Fall 2002), 21–33.

The Neuharth Solution

It was Gannett's Al Neuharth who found a solution to this problem. Under his guidance, Gannett accumulated monopoly newspapers in medium-size markets where the threat of competition was remote. Neuharth motivated his publishers to practice earnings management. They held earnings down during the good times by making capital investment, refurbishing the plant, and filling holes in the staff. And they boosted earnings in the bad times by postponing investment, shrinking the news hole, and reducing staff.

Gannett papers played Al's money game vigorously enough to produce a long period of steady quarter-to-quarter growth that satisfied the analysts' lust for predictability. Any long-term costs to these behind-the-scenes contortions did not bother them. Neither, for that matter, did the fact that some of the growth was unreal, because analysts and accountants alike are accustomed to looking at nominal dollar values rather than inflation-adjusted dollars. Neuharth's glory days were also a period of high inflation, and that helped to mask some of the cyclical twists and turns.

The price of Gannett stock soared. Managers of the other public companies saw what was happening and began to practice earnings management, too. One of the devices was the contingency budget, which was more like a decision tree than a planning tool. An editor is told how much he or she can spend on the news product in a given year provided that revenues remain at a certain level. If revenue falls below expectations, leaner budget plans are triggered at specified points on the downward slope.

Neuharth's showmanship worked just long enough to raise everyone's expectations about the value of newspapers. Today, despite some heroic efforts, not even Gannett can match Neuharth's record of "never a down quarter." Inflation is no longer high enough to mask the fluctuations in real return. Readers are drifting away. Advertisers are exploring other routes for their messages. In 1946, at the dawn of the age of television, newspapers had 34 percent of the advertising market.[4] In the second half of the twentieth century, newspaper share of the overall advertising market fell from almost 30 percent to close to 20 percent.

4. Jon G. Udell, *The Economics of the American Newspaper* (New York: Hastings House, 1978), 30.

Picard reminded us that, despite loss of share, newspapers still made more money than ever, primarily because the size of the advertising market grew, even after inflation was taken into account. However, it did not grow as much as gross domestic product. Newspapers, by raising prices and reducing cost, did well, but newspaper advertising as a share of GDP fell from seven-tenths of a percent to half a percent in the half-century.[5] Charging more for delivering less is not a strategy that can be carried into the indefinite future. So where will it all end? To envision the future, it helps to think about the readers.

The readership decline was first taken seriously in the late 1960s, when new information sources began to compete successfully for the time of the traditional newspaper reader. Competition spawned by technology began long before talk of the electronic information highway. Cheap computer typesetting and offset printing led to the explosive growth of specialized print products that could target desired audiences for advertisers. Low postal rates combined with cheap printing and computerized mailing lists spurred the growth of direct mail advertising. In short, the owners of the traditional toll road have been in trouble for some time.

Some observers draw a line on the chart of newspaper decline (see chapter 1), use a straightedge to extend it into the future, and foresee the death of newspapers. The reality is likely to be quite different. There is room for newspapers in the nonmonopoly environment of the newspaper future. They will not be as profitable, and that is a problem for their owners—whether they are private or public shareholders—but it is not a problem for society.

Imagine an economic environment in which newspapers earn the normal retail margin of 6 or 7 percent of revenues. As long as there are entrepreneurs willing to produce a socially useful product at that margin—and trust me, there will be—society will be served as well as it is now. Perhaps those entrepreneurs will not be the same ones who are serving us now, and that is not necessarily a concern to customers—except for one problem.

The problem is that there is no easy way to get from a newspaper industry used to 20 to 40 percent margins to one that is content with 6 or 7 percent. The present owners have those margins built into their

5. Picard, *Evolution,* 20.

expected return on investment, which is related to their standard of living.

It is return on investment that keeps supermarket owners content with 2 percent margins. And it is return on investment that makes newspaper owners, whether they be families, sole proprietors, or public shareholders, want to preserve their 20 to 40 percent. All of the money that they have sunk into the industry, whether by buying newspapers or spending on buildings and presses, has been cost-justified on the basis of that fat profit margin.

Look at it this way. If I sell you a goose that lays a golden egg every day, the price you pay me will be based on your expected return on investment (ROI), which needs to beat what the bank would pay on a certificate of deposit, but not by much. In negotiating the price that you are willing to pay me (and at which I am willing to sell), we'll both look for an expected ROI that compares favorably with other possible investments. And the reasonable assumption will be that the goose will continue to produce at the same rate.

Fast forward a bit. Once safe under your roof, the goose drops its production to one golden egg a week. That makes you a major loser.

Now, it's still a pretty good goose. You can resign yourself to the reduced income, or you can sell it to a third somebody who will be proud to own and house and feed it. And that new owner can, of course, get the return on investment that you were hoping to receive by simply paying one seventh of the price you paid.

What happened to the rest of the goose's value? I captured it when I sold it to you on the basis of the seven-day production schedule. The third owner is a winner, too, because he gets a fair return on his investment. Society is OK because there are plenty of other sources for golden eggs. The only loser is you.

Avoiding the fate of the second owner of the goose is the central problem facing newspaper owners today. They know they have to adjust to the reduced expectations that technology-driven change has brought them. They just don't know how. To understand the range of possible adjustments, consider two opposing scenarios laid out by business strategist Michael E. Porter:

Scenario 1. The present owners squeeze the goose to maintain profitability today without worrying about the long term. This is the take-the-money-

and-run strategy. Under this scenario, the owners raise prices and simultaneously try to save their way to profitability with the usual techniques: cutting news hole, reducing staff, peeling back circulation in remote or low-income areas of less interest to advertisers, postponing maintenance and capital improvement, holding salaries down.

It can work. A good newspaper, some sage once observed, is like a fine garden. It takes years of hard work to build and years of neglect to destroy. The advantage of the squeeze scenario for present-day managers is that it has a chance of being successful in preserving their accustomed standard of living for their career lifetimes. Both advertisers and readers are creatures of habit. They will keep paying their money and using the product for a long time after the original reasons for doing so have faded. When the bad end finally comes, the managers who locked the company into the strategy can say, "It didn't happen on our watch." Porter calls this strategy "harvesting market position."[6]

Scenario 2. The present owners—or their successors—will accept the realities of the new competition and invest in product improvements that fully exploit the power of print and make newspaper companies major players in an information marketplace that includes electronic delivery. That would be consistent with Porter's advice in chapter 1. "Rather than viewing a substitute as a threat, it may be better to view it as an opportunity," he says. "Entering the substitute industry may allow a firm to reap competitive advantages from interrelationships between a substitute and a product, such as common channels and buyers." The movement of newspaper companies into Internet distribution of news and advertising is a good example, because it exploits the newspaper's experience at creating content in the new distribution medium.

It was that prospect, in fact, that took my old outfit, Knight Ridder, into an experiment with electronic delivery of information way back in 1978. It offered hope of reducing the burden of high variable costs in the newspaper business. (A variable cost is one that increases with each unit of production, as opposed to fixed costs that are expressed in units of time. As circulation increases, the cost of newsprint, ink, and transportation rises in direct proportion. For an electronic distribution system,

6. Michael E. Porter, *Competitive Advantage: Creating and Sustaining Superior Performance* (New York: Free Press, 1985), 311.

the analogous costs are basically the same whether a hundred or a million consumers read your content.)

Under the second scenario, newspaper companies would build, not degrade, their editorial products. And there is a way to profit from the interrelationship between the old technology and the new. Tufts University political scientist Russell Neuman hints at it in "The Future of the Mass Audience." There is a way for newspapers to preserve at least some of their monopoly power. He calls it the "upstream strategy." Find another bottleneck further back in the production process.[7]

Originally, the natural newspaper monopoly was based on the heavy capital cost of starting a hot-type, letterpress newspaper operation. That high entry cost discouraged competitors from entering the market. Today, computers and cold type have made entry cost low, but the tendency toward one daily umbrella paper per market has continued unabated. That is because the source of the monopoly involves psychological as well as direct economic concerns. In their efforts to find one another, advertisers and their customers tend to gravitate toward the dominant medium in a market. One meeting place is enough. Neither wants to waste the time or the money exploring multiple information sources. This is why the winner in a competitive market can be decided by something as basic as the amount of classified advertising. One paper becomes the marketplace for real estate or used cars. Display advertisers follow in what, from the viewpoint of the losing publisher, seems a vicious cycle. From the viewpoint of the winner, of course, it is a virtuous cycle.

Neuman's thesis is that the competitive battle across a wide variety of media and delivery systems will make content the new bottleneck. "What is scarce," he says, echoing Herbert A. Simon, "is not the technical means of communication, but rather public attention." Getting that attention depends on content. He cites the victory of VHS over Betamax for home video players. Betamax had superior technology, but the buyers of VHS were attracted by the content because the manufacturer made sure that the video stores had VHS tapes.

How would that principle apply to newspapers? If the argument in

7. W. Russell Neuman, *The Future of the Mass Audience* (New York: Cambridge University Press, 1991), 150.

chapter 1 is correct, the most effective advertising medium is one that is trusted. If, as Hal Jurgensmeyer proposed, we define the newspaper's product not as information so much as influence, then we have an economic justification for editorial quality. The quickest way to gain influence is to become a trusted and reliable provider of information.

Trust as the Bottleneck

Trust, in a busy marketplace, lends itself to monopoly. If you find a doctor or a used car dealer that you trust, you'll keep going back without expending the effort or the risk to seek out alternatives. If Walter Cronkite is the most trusted man in America, there can be only one of him. Cathleen Black, when she headed the Newspaper Association of America, was getting at the same idea when she exhorted her members to capitalize on the existing "brand name" standing of newspapers. Brand identity is a tool for capturing trust.

And newspapers are in a good position to win that role of most trusted medium based on their historic roles in their communities. Under Scenario 2, they would define themselves not by the physical nature of the medium, but by the trust that they have built up. And they would expand that trust by improving services to readers, hiring more skilled writers and reporters, and taking leadership roles in fostering democratic debate.

Which scenario are we moving toward—harvesting the goose, or nurturing it and integrating it with new technology? The signals are mixed. During the new century's first recession, they tilted toward the harvest scenario. Reporters, once secure in their jobs, now hold what Herbert Gans has called "contingent employment."[8] When the *Duluth News-Tribune* discovered it was not meeting the year's profit goals set by parent Knight Ridder, it decided that it could meet its community service responsibilities with eight fewer reporters, and out the door they went. Layoffs, closing bureaus, and shrinking news holes became commonplace.

On the other hand, the public journalism movement represented an effort to build civic spirit in a way that would, if carried out over a long period of time, emotionally bind citizens to the newspaper. Whether

8. Herbert Gans, *Democracy and the News* (New York: Oxford University Press, 2003), 8.

very many newspapers will spend the money to wholeheartedly practice genuine public journalism remains to be seen. The harvesting scenario produces visible and immediate rewards while its costs are hidden and distant. The nurturing scenario yields immediate costs and distant benefits.

The dilemma cuts across all media and forms of newspaper ownership, but publicly held companies bear a special burden because of Wall Street's habit of basing value on short-term return. Take the case of Knight Ridder. With total average daily circulation of four million, its newspapers would bring a total of $7.2 billion if sold separately at an average value of $1,800 per paying reader. (McClatchy paid the Daniels family more than $2,400 per unit of circulation for Raleigh's *News & Observer*, but Raleigh is a better than average market). With 82.3 million shares outstanding at the early 2003 price of $64 per share, the entire company, including its nonnewspaper properties, was valued by its investors at only $5.3 billion, or $1.9 billion less than the break-up value.

How would a successful takeover bidder tap that potential $1.9 billion? By selling the papers to harvest-oriented publishers who would slash costs and build the bottom line with a bare-bones product. And how can public companies avoid such takeovers? One way is to do the harvesting themselves.

That's in the near term. Now stretch your time horizon beyond anything seen by Wall Street and imagine the final stages of the squeeze scenario. A newspaper that depends on customer habit to keep the dollars flowing while it raises prices and gives back progressively less in return has made a decision to liquidate. It is a slow liquidation and is not immediately visible because the asset that is being converted to cash is intangible—what the bean counters call "goodwill."

Goodwill is the organization's standing in its community. More specifically, it is the habit that members of the community have of giving it money. In accounting terms, it is the value of the company over and above its tangible assets such as printing presses, cameras, buildings, trucks, and inventories of paper and ink. I asked two people who appraise newspapers for a living, John Morton of Washington, D.C., a former analyst, and Lee Dirks of Santa Fe, a newspaper broker, to estimate the proportion of a typical newspaper's value represented by goodwill.

Both gave the same answer: 80 percent. That leaves only 20 percent for the physical assets.

This is vital intelligence for an entrepreneur interested in entering a market to challenge a fading newspaper. As an existing paper cuts back on its product and its standing in the community falls, there must come an inevitable magic moment when a competitor can move in, start a paper, build new goodwill from scratch, and end up owning a paper at only 20 percent or one-fifth of the cost of buying one.

Such a scenario is overly simplified, of course. The entry of competition could be just what it takes to get the existing paper to switch to a Scenario 2 strategy. But the newcomer would have a tremendous advantage, and that is its lower capitalization. Because its outlay is only the cost of the physical plant, one-fifth the value of the existing paper, the challenger can get the same return on investment with a 6 percent margin that the old paper's owners get with a 30 percent margin. Voilà! A happy publisher with a 6 percent margin! Because this publisher is building goodwill from scratch, he or she can cheerfully pour money into the editorial product, expand circulation, create new bureaus, heavy up the news hole, and do the polling and special public interest investigations that define public journalism.

This dream is not so wild. Remember Al Neuharth. One of the factors that propelled him to the top at Gannett was his astuteness in recognizing a parallel situation in east central Florida. Rapid population growth stimulated by space exploration had created a community that needed its own newspaper. He founded *Florida Today* for significantly less than the cost of buying an existing paper. The only obstacle is finding the right time and place—plus an opposition that is greedy and either shortsighted or slow-footed enough to continue squeezing out the old margins in the face of a challenge.

To old newspaper hands, the prospect of battles between the newspaper squeezers and the newspaper nurturers has a definite charm. Some of us old enough to remember the fun of working in competitive markets would line up to work for the nurturer against the squeezer. But the threat to companies that are liquidating their goodwill might come from another direction. It might not come from other newspaper companies at all.

The race to be the entity that becomes the institutional Walter Cronkite in any given market will not be confined to the suppliers of a particular delivery technology. How the information is moved—copper wire, cable, fiberglass, microwave, a boy on a bicycle—will not be nearly as important as the reputation of the creators of the content. Earning that reputation may require the creativity and the courage to try radical new techniques in the gathering, analysis, and presentation of news. It might require a radically different definition of the news provider's relationship to the community, as well as to First Amendment responsibilities.

It is fashionable to blame Wall Street for the bind in which newspaper companies find themselves. However, not all investors and analysts have a narrow, short-term orientation. Analyst O'Donnell, writing more than two decades ago, observed that quality journalism "can be expensive, but it helps a paper build an image that attracts talented employees both in news and other departments. We have spoken to employees in press rooms, for example, who take great pride in working on a newspaper that wins national awards . . . the perceived 'quality' of a paper can be a critical factor in morale and is not to be underestimated . . . readers' perceptions of the value of the product are substantially related to the quality of the news coverage."[9]

Analysts and investors make their money by spotting trends and taking investment positions in them before their competitors do. If journalistic quality is to have value on Wall Street, we will have to make the case that it is a leading indicator of profitability. If it is, savvy investors will find out eventually. The free market makes it inevitable. But the market sometimes takes a very long time to work its will, and we should not expect existing newspaper organizations to help very much. Their inherent conservatism, a consequence of their easy-money history, places them at a disadvantage in attempts at innovation. The pressures to harvest their market position by squeezing out the historic margins in the short term have made them inflexible. But if influence is the product, sooner or later some business entity will find a way to package and sell it, and the castle that it builds on its rock will shelter the best and the brightest creators of content.

---

9. O'Donnell, "The Business of Newspapers," 13–14. Like several other analysts, O'Donnell spoke from the vantage point of a former newspaper person. He was Knight Ridder's director of corporate relations in the late 1970s.

3

How Advertisers Make Decisions

I F   T H E influence model is a realistic description of how newspapers really work, there should be some evidence that advertisers consider a newspaper's influence when making their buying decisions.

The evidence is obvious when we compare extreme differences, such as *National Enquirer* versus *The Washington Post*. It gets harder to see when we look for subtle differences among mainstream newspapers. But change is afoot. By the turn of the twenty-first century, the advertising profession was in every bit as much turmoil as journalism—also as the result of new technology. While the turmoil was painful for all concerned, it left the business open to new ideas and ways of thinking.

The traditional way of evaluating advertising is by gross rating points, or "counting eyeballs." The industry has been ingeniously detailed in its efforts to develop specialized counts.[1] Probability models are used to estimate such esoterica as

1. For example, Kent M. Lancaster and Helen E. Katz, *Strategic Media Planning* (Lincolnwood, Ill.: NTC Business Books, 1989).

the number of unduplicated ad exposures in extremely narrow reader categories such as women ages 25–45 in a given market who purchased shoes in the previous six months.

Data for these quantitative measures typically come from just two sources. A newspaper's circulation, verified by the Audit Bureau of Circulations in most cases, tells the potential size of the audience. Survey research, conducted by large-scale national polls, gives a clue to the number who actually read or look into a given publication on an average day. It is the readership surveys that provide the basis for estimates within such narrow categories as the younger shoe-buying women.

Advertisers use a simple calculation called cost per thousand impressions (or CPM) to estimate the relative value of different media mixes. An "impression" simply means, in advertising jargon, that the message, whatever the medium, was delivered to the targeted consumer. It does not tell whether she looked at it (or, in the case of radio, listened to it), much less whether she acted on it.

Because advertisers like efficiency, they use survey data to try to minimize duplication. A national advertiser who wants to sell me a mail-order computer could get two chances by advertising in both *USA TODAY* and *The Wall Street Journal* because I am a home-delivery customer for both. But he or she is more interested in reaching two people once rather than one person twice, so I would be considered an inefficient duplication. The survey data helps ad buyers maximize net or unduplicated reach.

Another issue is frequency. While the number of daily newspaper readers has declined steadily since the 1960s (see chapter 1), the number of those who read at least once a week has held fairly constant. Therefore, repeating an ad several times during the week will bring in readers who missed earlier opportunities to see it. The outcome of higher frequency is an increase in net or unduplicated reach. Or, to put it crudely, more eyeballs. Michael Naples, who studied the issue of ad frequency enough to write a book about it, declared that three exposures to a target audience is optimal. One exposure, he said, has little or no effect; at two exposures you start to see some response; and efficiency peaks at three exposures. After that, there is still additional return but it's less likely to be worth the cost.[2]

---

2. Quoted in Jack Z. Sissors and Lincoln Bumba, *Advertising Media Planning* (Lincolnwood, Ill.: NTC Business Books, 1995), 140.

Influence gets into the picture in a rudimentary way. The sophisticated advertiser will use survey data to estimate the impact of a particular kind of ad by asking survey respondents if they remember the ad or can name the product. These numbers help establish general weights or rules of thumb that represent estimated impact. In one textbook example, a color ad gets more than twice the weight of the same ad in black and white. The back page of a newspaper section is better than the front (probably because of fewer distractions).[3]

But that's not the kind of influence we're talking about in the quest for an economic rationale for quality journalism. There ought to be a way to measure not just awareness but also the elements of trust and bonding with the community. An effective ad in a trusted publication would be worth more to advertisers.

Advertisers might be making such calculations intuitively, without realizing it. The marketplace absorbs information in mysterious ways. When the space shuttle *Challenger* crashed in 1986, it took a panel of experts several months to figure out that the defective component was the system of O-rings connecting segments of Morton Thiokol's solid rocket booster that was used to bring the space shuttle into orbit. But the stock market quickly sensed which of the four main contractors was responsible, and the knowledge was reflected in the value of Morton Thiokol shares within just a few days.[4]

The collective wisdom of markets was recognized by the Defense Department in 2003 when it tried to create a Policy Analysis Market to anticipate terror strikes before they happened—a good idea that was shot down by bad public relations.[5]

If Adam Smith's "invisible hand" of the marketplace is at work in setting advertising rates, then influential newspapers should be getting more for their advertising than those with less influence—whether they recognize the reason for it or not. We could test this if we had a good operational, that is, measurable, definition of influence. In chapter 1, we found that credibility made a pretty good surrogate for influence. To

3. Lancaster and Katz, *Strategic Media Planning*, 39.

4. Michael T. Maloney and J. Harold Mulherin, "The Complexity of Price Discovery in an Efficient Market: The Stock Market Reaction to the *Challenger* Crash," *Journal of Corporate Finance* 9:4 (September 2003), 453–79.

5. Peter Coy, "Betting on Terror: PR Disaster, Intriguing Idea," *BusinessWeek* (August 25, 2003), 41.

simplify the argument, let's just think of it as trustworthiness. I'm going to try to convince you that trusted newspapers are able to ask more for advertising than those that are less trusted. What follows now is the statistical case for that proposition.

Remember that the Knight Foundation study includes independent samples of five hundred persons in each of the twenty-six communities where John S. and James L. Knight operated newspapers in their lifetimes. The question it asked is, "Please rate how much you think you can believe each of the following news organizations I describe. First, the local daily newspaper you are most familiar with. Would you say you believe almost all of what it says, most of what it says, only some, or almost nothing of what it says?" For cross-market comparison of newspaper credibility, the percent who said they believed almost all of what the paper says makes a convenient benchmark.

Because the question is not specific to any particular newspaper, it measures newspaper credibility in the market as a whole. Therefore, it is necessary to limit this analysis to those markets that met three tests:

1. Market definition in the Knight Foundation survey was based on a whole county or combination of counties.[6]

2. Newspaper advertising in the county is dominated by a single newspaper or combination under joint ownership or management.

3. The circulation of the dominant newspaper is verified by the Audit Bureau of Circulations and does not exceed three hundred thousand.[7]

In the case of Fort Wayne, which has two newspapers with a joint operating agency including the advertising department, I used the com-

6. This eliminated Gary, Indiana, and Long Beach, California, where the survey areas were defined by zip codes. Two markets, Miami and Columbia, South Carolina, were defined by pairs of counties. In the Miami case, we split them and examined Miami-Dade and Broward counties separately. In South Carolina, we treated Lexington and Richland counties as a single unit.

7. Baldwin County, Georgia, was eliminated because its main paper's circulation is not audited. Markets eliminated by virtue of their size were Philadelphia, Detroit, and Miami. The source for circulation data was the electronic version of County Penetration Reports, Spring 2000. Audit Bureau of Circulations, Chicago. Most of the audits reported there were conducted in 1999.

bination advertising and circulation figures. (See Table 3–1 for a list of papers and their home county penetration.)

The third decision was made after exploratory analysis showed that variance in advertising rates is much greater for very large markets, which are susceptible to a greater variety of influences, than for smaller ones. Whether this is strictly a function of their size or due to peculiar circumstances in each large market, one cannot be sure. Either way, the theory doesn't work for the largest newspapers.

Those eliminated by virtue of size were *The Detroit Free Press, The Philadelphia Inquirer,* and *The Miami Herald.* In Philadelphia, where the *Inquirer* and the *Daily News* are about even in Philadelphia County, I could have kept the *Daily News* because it met the under three hundred thousand test for total circulation. But because its fate is so closely tied to that of the *Inquirer,* it seemed more consistent to leave it out. (The result doesn't change substantively.)

These decisions left a convenience sample of twenty-one newspapers whose circulation ranged from 289,814 (*San Jose Mercury News*) to 16,038 (*Aberdeen American News*). In addition, *The Union-Recorder* of Milledgeville, Georgia, was eliminated because its circulation is not verified by ABC. The two papers left out because the Knight Foundation survey did not include whole counties were the *Press-Telegram* of Long Beach, California, and the *Post-Tribune* of Gary, Indiana.

The Knight Foundation surveys, with a minimum sample size of five hundred, were conducted in 1999 and 2002. Because the single credibility question is subject to fluctuation, the two surveys were averaged for a better estimate of newspaper credibility in the twenty-one counties.

Analysis: Uncovering the Hidden Hand

The source for published advertising rates was the monthly report of Standard Rate and Data Service for April 2002.[8] The weekday rate per Standard Advertising Unit (SAU) was used if reported, otherwise the column-inch rate. The range was from $352 for the *South Florida Sun-Sentinel* (the Tribune Company's paper in Fort Lauderdale) to $29 for the *Aberdeen American News.* The rates are shown in Table 3–1.

8. SRDS Newspaper Advertising Source, 84:4 (Des Plaines, Ill., April 2002).

Table 3-1

City	ABC 2000 circulation	Home county penetration	EBI ($000,000)	Credibility score	Published SAU rate
San Jose	289814	40	47115	20.7	243
Ft. Lauderdale	257882	30	33618	22.4	352
Charlotte	245239	50	29077	18.3	221
St. Paul	200408	45	65336	18.4	145
Palm Beach	171619	34	31034	19.9	172
Akron	143050	47	13035	17.3	120
Columbia	119837	41	9411	20.8	126
Lexington	114275	48	8921	21.4	119
Ft. Wayne	107471	62	9327	19.5	91
Wichita	88973	39	10193	20.2	94
Macon	69076	49	4575	19.8	64
Tallahassee	51305	43	5059	16.2	70
Duluth	50589	40	3694	21.8	63
Biloxi	49489	48	5715	24.3	47
Columbus	48883	44	3921	24.3	58
Myrtle Beach	45443	51	3210	22.1	36
Bradenton	42076	39	12924	20.1	40
Grand Forks	34439	59	1550	26.5	36
Boulder	33119	29	7408	15.1	37
State College	24969	48	2159	20.8	33
Aberdeen	16038	64	705	27.2	29

The obvious source for such wide variation in published advertising rates is circulation size. It alone accounts for 88 percent of the variance in posted ad prices.

Some of the remaining variance can be explained by the wealth in the market, a value estimated by SRDS and reported at the county level as Effective Buying Income (EBI). Home county circulation penetration is also a factor. A newspaper whose market is compact and contiguous is more likely to be useful to local retail advertisers, and a relatively high home county penetration distinguishes such a market from one whose circulation is thinly spread. The range for home county penetration was from 30 percent (*South Florida Sun-Sentinel*) to 64 percent (*Aberdeen American News*).

The hypothesis that citizen trust, as measured by the Knight Foundation question, can predict published advertising rates was tested with a variation on the idea of correlation. It is a multiple regression model, and it estimates the effect of individual variables after the effects of the others have been taken into account.

To keep it simple, the following table just shows the good stuff. Each line shows how much variation in ad rates is predicted with each additional variable, starting with circulation. The table is cumulative. Thus, circulation alone explains 88.2 percent of the variation in ad rates, and adding EBI to the equation boosts that two percentage points, to 90.4. The third column gives the overall significance of each variable's contribution to published ad rates.

After circulation, home county penetration and credibility make the more meaningful contributions. (The second column is headed by the statistical term for "variance explained" or R^2 because it takes up less room.) The column for significance gives the statistical probability that the effect is a chance result of the small sample.

Variable	R^2	Significance
Circulation	.882	.000
EBI	.904	.028
Penetration	.926	.003
Credibility	.947	.023

The hypothesis is supported. Credibility explains an additional 2.1 percent of the variance after circulation, home county penetration, and EBI have all had their effects accounted for.

For a better visualization of what is happening, the regression was rerun, leaving out the credibility variable and saving the unstandardized residuals. These numbers represent the variability in advertising rates that is left over after making adjustments for circulation, home county penetration, and EBI. This adjusted SAU rate ranged from 84 in the case of Fort Lauderdale's *Sun-Sentinel* to -32 for the *Saint Paul Pioneer Press*. In other words, the *Sun-Sentinel* asks $84 more per SAU than its circulation, home county penetration and EBI would justify, while the *Pioneer Press* asks $38 less.

The scatterplot below shows the effect of credibility on adjusted ad rates and identifies the main outliers.

Figure 3-1

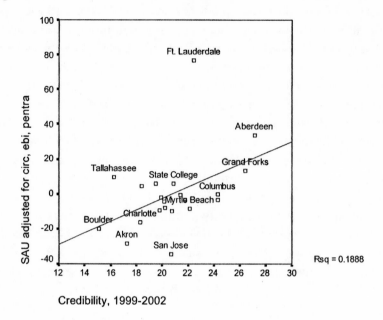

Credibility, 1999-2002

Credibility explains 19 percent of the residual variance in published advertising rates. Its effect is statistically significant.[9]

The slope indicates that, other factors being equal, each percentage point increase in credibility is worth an additional $3.25 in average SAU rate. Please remember that this is a limited sample, heavily influenced by that outlier at the top, the *South Florida Sun-Sentinel,* which is above average in credibility and very high in published advertising rate. Many more cases are needed to make this result generalizable beyond the Knight communities.

At a meeting in New York City in 1985 to discuss that year's ASNE credibility study, David Weaver suggested a need to find out "how much credibility is worth and how much it costs to get it." We now have a tentative answer to the first question. Based on the mean ad rate in this group of twenty-one newspapers of $131.21, this result suggests that,

9. R = .435, p = .049. An earlier version of this analysis without the 2002 Knight Foundation survey was presented by Joe Bob Hester and me in "Trust and the Value of Advertising: A Test of the Influence Model," American Association for Public Opinion Research, Nashville, May 17, 2003.

other factors being equal, a one point improvement in credibility is worth a 2.5 percent increase in a newspaper's asking price for advertising.

But hold on. It's only an asking price. Newspaper rates are fixed more firmly than rates for broadcast media, where buying local advertising is a lot like buying a car and few pay the sticker price. Broadcasters' rate cards, in the words of one textbook, are "simply a starting point for negotiation."[10] Newspapers, although more stable than that, sometimes have different rate cards for different classes of advertisers. Some will include value-added elements in the price such as creative services or access to audience research. An ad that is repeated might be priced at a lower "pick-up" rate because the newspaper's cost of composing it has already been covered. Separate advertising and editorial content for different geographic zones further complicates the price issue because it makes it possible to buy only certain desired geographic segments of a newspaper's circulation.

Steve Rossi, president of Knight Ridder's newspaper division, was present when I presented these results at a Poynter Institute gathering, and he pointed out the problem. As a friend will, he accompanied his criticism with a remedy. Eighteen of the papers in the sample were members of the Knight Ridder group, and Rossi was able to provide the mean advertising rates actually charged by those papers.[11] It turned out that the published price is highly correlated with actual price, explaining 90 percent of its variation.[12] But that doesn't mean a lot because it is the variance after circulation is accounted for that interests us. Here the correlation is still positive. When published rate and actual rate are expressed as dollars per thousand circulation, the published rate explains most of the variation (60 percent) in actual rates.[13]

So published rates are not meaningless. While that improves confidence in their validity, the next step does not. When the four-variable regression is run with the eighteen Knight Ridder papers, using the actual ad price as the dependent variable, nothing has predictive power

10. J. Thomas Russell and W. Ronald Lane, *Kleppner's Advertising Procedure* (Upper Saddle River, N.J.: Prentice Hall, 1999), 273.

11. My thanks to Virginia Dodge Fielder, Knight Ridder Vice President/Research for assisting with this task.

12. $R = .951$.

13. $R = .774$, $p < .0002$.

except the absolute value of total circulation. All of the interesting effects, including home county penetration, effective buying income, and credibility, simply vanish.

This makes us suspect that the price/credibility effect is driven by the three non-KR papers in the sample (Boulder, Palm Beach, and Fort Lauderdale). It's easy to check, and you don't need statistical tests. Just take another look at the chart in Figure 3–1. Boulder is down in the low-rate, low-credibility corner, and Fort Lauderdale soars in the opposite corner. Palm Beach is one of those unlabeled points in between. Their presence is creating the whole effect. That is another reason to look beyond this small and peculiar sample.

Much more data collection is needed. A better measure of influence would help, too. As a minimum, there ought to be a multi-item measure of newspaper credibility like the one used by Chris McGrath and Cecilie Gaziano in their 1985 credibility study for the American Society of Newspaper Editors. (I'll explain why I like it in chapter 4.)

The Knight Foundation's single item will become more powerful as more data points are accumulated. If the pattern of a survey every three years continues for another thirty years, there will be twelve time points to compare, and a clearer picture could emerge—if newspapers, as we know them, still exist by then.

Meanwhile, I choose to believe that something real has surfaced, briefly and seductively, in this data set. Somebody with more time and better data ought to try to replicate it. Meanwhile, let us turn to other signs that influence is a real or potentially real factor in determining the value of advertising.

The Search for More Precision

The most encouraging news is that CPM—the concept created to help ad buyers find the most eyeballs at the least cost—is falling out of favor. It might still be popular with the sell-side analysts of Wall Street, who like it for its simplicity, but it gets less and less attention among the real decision makers. Part of the problem is the imprecision of the number, based as it is on tiny samples within each narrow demographic group that advertisers care about. Questions were raised as far back as the mid 1990s.

"Some marketing and media planners believe that the cost-per-thousand concept is a dead issue," said Sissors and Bumba in their 1995 text. ". . . just because one vehicle has a lower CPM than another does not mean that it has more value than another. Many planners and marketing executives believe CPM measures value. But CPM is only remotely related to sales. Lower CPM does not mean more value."[14]

Those planners and marketing executives represent the sophisticated end of the ad buying business. At the other end, small newspapers, decision makers seldom had enough data to worry about CPM even if they wanted to. Ken Smith, a former newspaper publisher turned professor at the University of Wyoming in Laramie, surveyed advertisers in six small cities to learn the basis of their ad buying decisions. Cost was seldom a factor, nor did the advertisers show much knowledge of audience sizes.

"Instead, their decisions are much more subjective in nature," reported Smith, "often based on personal or emotional rather than business factors."[15]

A large number base their evaluation on little more than word of mouth. A customer walks in the store and says, "I saw your ad." Or the advertiser might notice that sales go up after an ad runs, but have no way to discount other factors such as weather, season, or employment rates. Much greater precision is attained when an ad contains coupons that are returned and counted.

We should not leap to the conclusion that the big-city advertising buyers with their hair-splitting CPM calculations are smarter than the grassroots guys. By being a part of the local culture and knowing the customers personally, the small-town retailer might be more likely than a national advertiser to recognize such a hard-to-measure factor as an advertising medium's influence.

Supporting evidence is found in another academic survey from the 1990s, which sampled advertisers in two college towns, Athens, Georgia, and Madison, Wisconsin. In response to an open-ended question, a fourth of the advertisers in both markets volunteered opinions that

14. Sissors and Bumba, *Advertising Media Planning,* 76–77.
15. Ken Smith, "Advertisers' Media Selection in Small Newspaper Markets," *Newspaper Research Journal* 19:1 (Winter 1998), 39.

"higher editorial quality and better production capabilities would increase daily newspapers value as an ad medium."[16]

One advantage that local advertisers have is their dual relationship with local media. On the job, the newspaper is their business tool, but it is also their off-the-job personal instrument, as Jana Frederick-Collins noted in a widely cited paper written while she was a graduate student at Chapel Hill.[17] This duality is a powerful force. When *USA TODAY* was struggling for recognition from the national advertisers on which it would ultimately depend, it established an advertising industry beat in its business section. That strategy helped make the paper a personal instrument for the people who make national ad buying decisions and probably shortened the long wait for profitability.

So we have some support for the following working hypothesis: influence is an existing factor in local ad buying decisions and a potential factor for national buyers. A brief look at the recent history of national advertising will show why I think the second part of that proposition is true.

Advertising and journalism used to have a lot in common. Both attracted bright, creative people who were there at least in part for the fun of it. The secondary reward for those who succeeded in advertising was lots of money, while for those who were good at journalism the rewards were mostly intangible, including the chance to influence public policy for the better.

Both businesses were affected by the late-twentieth-century move toward consolidation, bureaucratization, and pressures for short-term profitability.

"For most of the last half of the 20th century, advertising was exciting, and it was fun," wrote the former publisher of *Advertising Age* in a contemporary history of the business. But the fun started to go out of it when the consolidation movement gained headway. "Within only a few years," said Joe Cappo, "the advertising agency business in the United

16. Glen J. Nowak, Glen T. Cameron, and Dean M. Krugman, "How Local Advertisers Choose and Use Advertising Media," *Journal of Advertising Research* 33:6 (November/December 1993), 46.

17. Jana Frederick-Collins, "Measuring Media Image: Expectations, Perceptions and Attitudes of Local Retail Advertisers," prepared for Association for Education in Journalism and Mass Communication, Montreal, August 1992.

States has transformed from dozens of independent, entrepreneurial, creative, and highly competitive shops into an oligopoly of four large publicly held corporations."[18] The four, Publicis Groupe, Omnicom Group, Interpublic Group of Companies, and WPP Group, accounted for 55 percent of all global advertising and marketing expenditures, Cappo estimated in 2003.

Creativity, formerly channeled into producing cute, clever, and emotion-arousing messages, was broadened to take in a wider array of decisions. The more important creative thinkers were those who could look at the total marketing problem and develop a strategy for deploying the best combination of tools. Advertising became integrated with sales promotion and public relations.

The way advertising was paid for began to change. When television was the main national medium, agencies charged a straight 15 percent commission, and that was enough to support plenty of auxiliary services such as audience research and copy writing at no extra charge. But as technology made the problem more complicated, agencies were forced to unbundle those services. At the same time, a unified strategy was needed to break through the noisy buzz of the information age and get messages into places where they would change behavior. Just as newspaper editors were realizing that they had to get their messages not just into people's hands, but into their heads, advertisers started worrying about how much their messages affected buying behavior.

One result was talk of an "involvement index" to be added to the standard audience measurements to help advertisers "reliably reach interested consumers when and where they're willing to engage."[19] If advertisers are willing to think about an involvement index, can an "influence scale" be far behind?

Newspaper publishers and their advertisers have been slow to pick up on new technological opportunities for measuring the degree to

18. Joe Cappo, *The Future of Advertising: New Media, New Clients, New Consumers in the Post-Television Age* (New York: McGraw Hill, 2003), 11.

19. Don Rossi, "New Media 'Involvement Index' May Change Forty-Year-Old Marketing Rules," *Advertising Age* 73:17 (April 29, 2002), 16. Reprinted in Cappo, *The Future of Advertising,* 54–57. See also Bradley Johnson, "Cracks in the Foundation: Why the Very Currencies the Industry Depends On Are Dated and Inadequate," *Advertising Age* 74:49 (December 8, 2003), 1.

which newspaper advertising affects sales. Bar code scanners, developed with the assistance of the National Science Foundation and first applied at a grocery store in Troy, Ohio, in 1974, made it possible to test the effect of product-specific advertising and special sales events with unprecedented speed and precision, but the newspaper industry was slow to exploit that opportunity.

When B. Stuart Tolley was director of advertising research for the Newspaper Association of America, he ran a nice demonstration project. Using Richmond, Virginia, newspapers, he designed a split-run experiment so that half the area got one set of ads and the other half a different set. All were image ads for national grocery products, including Nabisco Fruit Newtons and Meow Mix. Purchasing was monitored through a local supermarket chain that captured both the purchase and the shopper's identity (through a Valued Shopper Plan card) at checkout. Those data, collected for twenty-five weeks, were merged with the newspaper subscription list so that Tolley could see which shoppers had been exposed to which ads. The Nabisco ad produced the strongest result, generating $4.40 in additional sales per ad dollar spent.[20] For a national brand trying to increase market share, that is significant.

The desire for measurement is one of the factors pushing advertising toward more direct marketing where effects are instantly and precisely recorded. Coupon ads in newspapers, the catalogs that fill up your mailbox, and those hated suppertime telephone calls are the familiar forms of direct marketing. They persist because their cost effectiveness is never a mystery.

New Forms of Direct Marketing

Now the Internet is bringing new forms of direct marketing. Like all technology, it can be used for good or for ill. Users range from Nigerian con artists promising to make you rich in exchange for access to your bank account, to Amazon.com recommending a book expressly for you based on its knowledge of your tastes as revealed in previous purchases.

Internet advertisers have the potential to establish their own influ-

20. B. Stuart Tolley, "A Study of National Advertising's Payout: Image Ads in Newspaper ROP," *Journal of Advertising Research* 33:5 (September–October 1993), 11–20.

ence. When the battery for my off-brand cell phone wore out, I went to the retailer who sold it to me and found that he no longer supported that brand and wasn't a bit apologetic about it. None of the other telephone or electronic retailers who advertised in my local newspaper carried the battery. I called the manufacturer, who was happy to sell me the batteries, but only in gross lots. On a hunch, I turned to Amazon.com, typed in the name of my off-brand phone, found it and the battery listed, and had my replacement battery two days later.

My cell-phone story is not entirely a happy one because it shows the ease with which technology can separate advertising from its traditional role of supporting socially useful editorial content. The Internet can be a perpetual catalog; it never clutters my mailbox, yet I can always find it when I want it.

The catalog function has been one of newspapers' strengths in competing with television. A newspaper is a pretty good information retrieval machine because you can interact with it by turning the pages to find specific product information. TV has prettier pictures, but it keeps you passive, and it lacks the bandwidth to give you very detailed product information. The Internet solves both of these problems.

As noted in chapter 1, a standard strategy for an industry disrupted by new technology is to focus on a function of its business that it can still perform better than the substitute. Newspapers, because of their editorial content and not in spite of it, are positioned to wield more influence than any of the substitutes. As community bulletin boards and suppliers of the policy-related information that is required for democratic deliberation, they have influence that would be very difficult for an Internet-based medium to duplicate. As a defensive measure, newspapers are rightly getting Web sites of their own, although they vary considerably in their dedication to taking full advantage of the value-added potential of the new medium. The most obvious application is classified advertising because of the speed with which ads can be posted, revised, and withdrawn as sales are made.

What's not clear is whether an Internet classified ad site must dominate its market in order to survive. The history of newspapers has shown that a given community will, in the long run, support only one marketplace. Buyers and sellers alike tend to converge on the spot where they are most likely to find one another, and the surviving papers in formerly competitive markets were generally those that had established themselves as the dominant classified ad medium.

But the Internet makes it easier to hop from one market to another. If I don't find the book I'm looking for on Amazon.com, I have no hesitation about going to Barnes and Noble, Borders, or Powell's, each only a mouse-click away. Market-hopping is so easy, in fact, that maybe the concept of a fixed location—albeit a hyperspace location—for a market is wrong-headed. The success of Google, as of this writing the most popular online search engine, makes us wonder. Consider the following observation by reporters for *The Wall Street Journal:*

> Google long ago realized something that is only dawning on many other companies: Searching isn't a Web sideline—it's the Web's strategic heart. While Amazon and other sites try to position themselves as the central place for online shopping, thousands of shoppers are simply Googling for sandals or curtains and whatever else they want.[21]

Google makes money off this behavior by selling pop-up ads that are related to the user's search and appear on the same page as the initial result of the search. What we have here is direct marketing with fiendishly clever efficiency, far better than a newspaper coupon.

Still, a newspaper beats the Internet in portability. You can read it in the bathroom, at the breakfast table, in the back yard, or on the bus. But computers are getting smaller and more portable and are less often in need of being hard-wired to the Internet. In the coffee houses on Franklin Street in Chapel Hill, I still see students reading newspapers. But they are being joined in increasing numbers by peers browsing the Internet on their wireless-connected laptops. Even cell phones are mutating into multiple-use devices that provide e-mail and Internet browsing.

These developments tend to leave very little for a newspaper to do better than anyone else—except for the influence factor. The newspaper's influence within its community, a good that can be created and sustained only by high-quality, geography-specific editorial content, will be very difficult for any new medium to replicate.

But it can be done. As I will remind you repeatedly, newspapers suffer from the historic fact that they are a manufacturing enterprise. They

21. Mylene Mangalindan, Nick Wingfield, and Robert A. Guth, "Rising Clout of Google Prompts Rush by Internet Rivals to Adapt," *Wall Street Journal,* July 16, 2003.

buy raw materials, ink and paper, add value with news and information and advertising, then transport and sell the finished product. The trouble with a manufacturing business is that its main costs are variable, not fixed. What this means is that every new customer means a proportional added cost in raw materials and transportation. Double your customers, double the newsprint.

The Internet, like broadcasting, is mostly fixed cost. If the number of searchers accessing a Web page increases tenfold, the cost of maintaining that page does not, within certain limits, change. Broadcasters have always had that advantage over newspapers. If the new media barons of the future make large investments in creating trusted editorial products that will attract and influence citizens and buyers, newspapers will be in trouble, but society will still be served.

Here's an example that journalists can appreciate. The Poynter Institute for Media Studies, the nonprofit institution that owns the *St. Petersburg Times,* wants to improve journalism by providing midcareer education to journalists. One of its education tools is its Web site. But how to attract viewers? It was, recalled Jim Naughton, former head of the institute, "rich with academic material that no one knew was there."[22]

Naughton and his marketing director/online editor, Bill Mitchell, solved the problem with a friendly takeover of Jim Romenesko's Web operation. Romenesko was an entrepreneur who rounded up news and gossip about the media from a variety of sources and posted it for the benefit of practitioners. Journalists spend a lot of time in front of computer terminals, and many got into the habit of taking breaks to see what Romenesko was dishing out about their peers. After joining Poynter, Romenesko performed the same reliable service, but you had to get to him by accessing Poynter first. It introduced a lot of people to Poynter and made many of us psychologically dependent on it.

Before Romenesko came aboard in 2000, the Poynter site was generating around 10,000 page views per weekday. By 2003, it was up to 160,000 a day, with spikes to 250,000 when there was a big story such as the departure of editor Howell Raines from *The New York Times.*[23] As a journalism teacher, I had, in that hot summer, more than a passing interest in following the *Times's* handling of its mendacious reporter Jayson

22. Telephone interview, August 14, 2003.
23. E-mail from Jim Naughton, August 20, 2003.

Blair. I could have used Google, but it was much more efficient to go straight to the Poynter site and get, in addition to Romenesko's roundup, the full array of Poynter faculty, contributors, and editors who had something to say on the subject.

I was not alone. By 2002, PoynterOnline was the trade publication most read by journalists, beating both the slick-paper publications, *American Journalism Review* and *Columbia Journalism Review*.[24]

The link between advertising and editorial content works in exactly that way. And while newspapers still enjoyed domination on the editorial influence side of the equation at the turn of the century, they stood to lose it to bolder and more farsighted entrepreneurs in new media if they were not careful.

Without being partial to one medium or another, researchers, whether in industry or in academe, ought to be vitally interested in this issue and strive to do two things:

1. Find a way to measure a medium's influence.
2. Establish a clearer relationship between its influence and the value of its advertising.

If we can't do that—or won't—it will soon be time to think about vehicles other than advertising-supported media to fulfill the social responsibility functions that newspapers have historically provided. Without such data, the shareholders in newspaper companies will continue to press for what amounts to a harvesting strategy, short-sightedly reducing the newspaper to a bare-bones advertising platform with neither influence nor prospect of survival. We should not let them give up so easily.

24. David H. Weaver, Randal A. Beam, Bonnie Brownlee, Paul S. Voakes, and G. Cleveland Wilhoit, "The American Journalist in the 21st Century," Association for Education in Journalism and Mass Communication, Kansas City, Missouri, August 1, 2003.

4

Credibility and Influence

W H E N I left graduate school at the age of twenty-seven and started covering education for *The Miami Herald*, a school official gave me some advice. "The *Herald* carries a big stick in this community," he said. "Use it carefully."

It didn't take long to figure out what he meant. The *Herald* was the fourth newspaper to hire me but the first to push me to practice journalism at the level described by Curtis MacDougall in his classic 1938 textbook, *Interpretative Reporting.*[1] My previous newspaper employers had been smaller, too dependent on the local power structure or too thin on resources to do much more than cover routine meetings and rewrite handouts. But Miami was a community of highly diffuse power, populated by people who had been socialized somewhere else, and the *Herald* was a centralizing and stabilizing force. Major local issues were debated in its lively pages every day. On election day, some citizens carried the *Herald's* editorial page into the booth so they could support or oppose its

1. Curtis D. MacDougall, *Interpretative Reporting,* rev. ed. (New York: MacMillan Co., 1948), was my introduction to reporting when I was an undergraduate at Kansas State.

endorsements. The "big stick" my school board source had described was the newspaper's considerable influence in that busy and conflict-ridden society.

If influence is a factor in determining the economic success of a newspaper, we need to learn how to define it, identify its parts, and measure it. Credibility is probably not the only component of influence, but it's a good place to start. Some strong efforts have been made to measure credibility, including the 1985 study "Newspaper Credibility: Building Reader Trust" for the American Society of Newspaper Editors by Chris McGrath and Cecilie Gaziano.[2] They surveyed newspaper readers in 1984 using a mailed questionnaire. It asked the readers to rate their newspaper on sixteen different criteria, using a five-point scale with a positive description at one end and its opposite at the other, for example, "fair—unfair." It included direct and unambiguous measures of believability, such as "accurate," "unbiased," "can be trusted," plus some other dimensions including the newspaper's relationship with its community. The latter group included "watches out after your interests" and "concerned about the community's well-being."

Gaziano and McGrath archived their data for secondary analysis. Later work has made a case for treating believability and community connectedness as separate dimensions.[3] In my own reanalysis, I proposed using five of their items in an index to measure believability:

Fair	Unfair
Unbiased	Biased
Tells the whole story	Doesn't tell the whole story
Accurate	Inaccurate
Can be trusted	Can't be trusted

And a four-item index of community affiliation:

Patriotic	Unpatriotic
Concerned about the community's well-being	Not concerned about the community's well-being

2. Published by the American Society of Newspaper Editors, April 1985.

3. Mark D. West, "Validating a Scale for the Measurement of Credibility: A Covariance Structure Modeling Approach," *Journalism Quarterly* 71:1 (1994), 159–68.

| Concerned mainly about the public interest | Concerned mainly about making profits |
| Watches out after your interests | Does not watch out after your interests. |

The overall 1–5 scales, believability in the first instance and community affiliation in the second, are obtained by simply averaging the individual items.

What you get are measures with strong internal consistency and a measure that is freer from error than if you just used a single question—as I did with the Knight Foundation data in chapter 1.[4] Multiple measures also give you a sense of what readers mean when they say a newspaper is credible. Part of what they mean is that they just like it. You can think of their attitude as a warm, fuzzy feeling toward the paper.

Evidence of this warm fuzziness is found in the fact that the two conceptually different measures, trust and community affiliation, are modestly related. They share 41 percent of their variance.[5] This common variance suggests that some underlying and unmeasured factor influences both of them. I choose to call this hidden factor "warm fuzziness" for now, and it is probably a good thing for newspapers to have, and it very likely has something to do with their influence. Social scientists have a name for these phenomena whose effects are evident without their being measured directly. They call them "latent variables."

The concept of latent variables is especially useful for people trying to measure newspaper quality. "I know it when I see it," editors have told me, "but I don't see how you can put a number to it." Maybe, maybe not. But if we can agree on enough interesting elements of quality that are measurable, and if there is statistical evidence that they are driven by some common underlying force not directly measured, we can make a good claim that the underlying force, even though it might be latent, is in fact quality. The beauty of this approach is that we can have our measures both ways. We can treat believability and affiliation

4. Internal consistency (reliability) is measured by Chronbach's Alpha. In my analysis of the Gaziano-McGrath data, it was .83 for believability and .72 for community affiliation. While there is no formal minimum standard, the existing literature commonly accepts .6 as adequate for exploratory analysis and .8 as the minimum for confirmatory work.

5. Pearson's r = .64.

as separate entities when it suits our purpose and combine them when the underlying latent variable is of more interest to us. An example from real life will make all this clearer.

When the world was younger and James K. Batten (1936–1995) was the vice president for news of Knight Ridder, one of his children asked, "Dad, why isn't there anything for kids in the paper?"[6] Since even back then in the 1970s we knew there was slow but steady readership loss and that it was at least partly due to generational replacement—older, loyal readers dying off and being replaced by young people socialized to other media—we could see that specific attention to content for kids might be helpful. I was assigned to find potential suppliers of children's content and work out a plan to recruit one to produce a daily newspaper feature. Networking my way around New York City's publishing culture, I found four brilliant creators of content for kids and persuaded each to turn out a prototype series that I then tested on groups of school children.

The winner was Scholastic, Inc., with a package called The Dynamite Kids' Page created by the same writers and artists that produced *Dynamite* magazine. We tested it in several markets, but chose the *Akron Beacon Journal* for the most formal and careful trial. Before the daily half page started on May 1, 1978, we measured 540 households where children were present, getting baselines for children's use of the paper and adults' feelings of credibility and general attitude toward the paper—the warm-fuzzy factor.

It was a simple pre-test, post-test design. If you have ever tried this, you know why Samuel Stouffer called it "a wide open gate through which other uncontrolled variables can march."[7] Even if you show the needle moving, you aren't sure why it moved. Anything that happens between the first and second tests might be the cause. You hope that nothing significant happens other than your experimental intervention, in this case the Dynamite Kids' Page. We weren't so lucky, but the unexpected event did bring us a windfall of insight into newspaper influence and credibility.

The first-wave measurement was based on interviews from March 28

6. James K. Batten, personal conversation, Washington, D.C., ca. 1977.
7. Samuel Stouffer, "Some Observations on Study Design," *American Journal of Sociology* 55:4 (January 1950), 357.

to April 8. On April 1, the *Beacon Journal* published information uncovered by its Washington correspondent David Hess about safety problems with a locally manufactured product, the Firestone 500 radial tire. That the newspaper was the initiator of this exposé didn't fully soak in until April 11, two days after our interviewing ended, when Firestone subpoenaed Hess in an effort to find out who had leaked the damaging data to him. The story was faithfully reported in the *Beacon Journal*. Before the month was over, Firestone began dumping its unsafe tires at discount prices in Florida and Alabama, a congressional investigation was opened, and Hess was grilled by Firestone and government attorneys trying to force him to reveal his source. He didn't.

Before the second wave of interviewing, June 17–30, the Firestone story just kept getting worse. The congressional probe was expanded to include another Akron tire maker (Goodyear), and Firestone reported a second-quarter loss and then announced that it would tear down one of its Akron plants with no plans to rebuild. The town was in trouble, and it was all too easy to blame the bearer of the bad news, the newspaper. So the experiment inadvertently became a measure of two things: changes in the use of the newspaper by children, and the newspaper's ability to maintain its influence while generating controversy.

The kids' page got fine reviews. The newspaper's rating as believable held steady, but it slipped on the warm-fuzziness items. On the responsible-irresponsible scale, it moved significantly toward the bad side. On the scale indicating whether it was or was not worth the price, it slipped in the direction of not worth the price.[8]

Throughout the discussion on credibility, some editors have wondered if the underlying variable is really lovability and whether a newspaper that tries too much to be loved risks losing its journalistic integrity and its ability to tell hard truths that the community needs to hear for its own good. There is an answer in Akron if we fast-forward to 1983, five years after the Firestone story.

A new survey, made for a different purpose by Urban and Associates,

8. Technical details of this case are reported in Philip Meyer, "Defining and Measuring Credibility of Newspapers: Developing an Index," *Journalism Quarterly* 65:3 (Fall 1988), 567. Although the Dynamite Kids' Page was a popular and artistic success, it was scrapped after Knight Ridder's national sales staff failed, despite much effort, to find child-directed advertising to support it.

found that the *Beacon Journal's* daily readers, by a margin of 77 percent to 7 percent, thought the newspaper was supportive rather than critical of its community. But the same daily readers considered its reporting biased rather than objective, by 48 percent to 37 percent. The two studies five years apart used different methods and different populations—households with children in the first instance, daily readers in the second. But it seemed clear that the weakness in public trust related to the Firestone story had dissipated after five years and even turned into a strength. On the other hand, its pure believability, strong five years earlier, showed signs of slipping.[9]

The Akron case suggests that influence and its two components, believability and community affiliation, can bend with the news and then recover. Editors who try to weigh the effect of every daily decision on the basis of how it would affect credibility and/or influence would be making a mistake. Telling the community news that it does not want to hear but nevertheless needs for its own long-term good could be the best way to maximize influence in the long run.

The Stability of Credibility

We can assess the stability of newspaper believability by looking at the Knight Foundation's twenty-six communities of interest (expanded to twenty-seven by splitting South Florida into Dade and Broward counties) and comparing their surveys of 1999 and 2002. As expected, belief in 1999 in a given community is a predictor of what belief will be in 2002. But, surprisingly, as noted in chapter 1, it is a weak predictor. The degree of shared variance is only 21 percent. In other words, if you tried to predict a community's trust in newspapers in 2002 based on its level of trust three years earlier, your prediction would be only 21 percent more accurate than if you had just assigned each community the overall average for 2002.

Across the twenty-seven communities, the percent who say they believe the newspaper they read most often all the time or almost all of the time was quite stable. The 1999 average, 21 percent, faded by less than half a percentage point by 2002. But individual communities fluc-

9. Described in Meyer, "Defining and Measuring Credibility of Newspapers."

tuated considerably. Sedgwick County, Kansas, gained seven points while Broward County, Florida, Grand Forks County, North Dakota, and Philadelphia County, Pennsylvania, all fell by more than five.

Some of this change can be explained by sampling error, but not nearly all. More is going on. Curious, I called Mike Jacobs, editor of the *Grand Forks Herald,* and asked if he had any idea why the percent who believe their favorite newspaper slipped from 29 percent to 23 percent in three years. He had a ready theory: the newspaper had endorsed a Native American proposal to drop "Fighting Sioux" as the nickname of the University of North Dakota's sports teams. It was a classic example of a newspaper's being a bit ahead of its community. Another possibility is that the paper's reputation was inflated by its Pulitzer-winning coverage of the floods on the Red River of the North in 1997 and that it reverted to normal by the time of the second survey.

In Wichita, the *Eagle* had been a step ahead of its highly conservative community in the 1990s with its aggressive reporting of controversial issues including the death penalty and abortion. That cost it some credibility, but it was only a temporary cost.

The mean absolute shift in credibility among the Knight Foundation communities was 2.7 percentage points with considerable variation around that mean. This volatility suggests that credibility is something like blood pressure: it needs to be measured more than once.

Think of credibility as having two components. One is a solid inner core that doesn't change from day to day or even from year to year. The other is the variable outer shell that is subject to the shifting winds of public mood as the news changes. This interpretation is consistent with the argument by William Schneider and Seymour Martin Lipset that confidence in institutions is pretty much a characteristic shared by institutions in general. When the public has confidence in the press, it also has confidence in the executive branch of the federal government (the Watergate years being the standout exception).[10] If we generalize from that to the case of newspapers, we begin to suspect that trust might not be so much a characteristic of a newspaper as of its community— or, perhaps more precisely, of the interaction between a newspaper and its community.

10. Seymour Martin Lipset and William Schneider, *The Confidence Gap: Business, Labor and Government in the Public Mind* (Baltimore: Johns Hopkins University Press, 1987).

How to measure the solid inner core? By taking repeated readings over time and smoothing them. A smoothed time series uses moving averages to let you see the basic trend by removing the peaks and valleys caused by one-time local conditions. As more Knight Foundation data is collected over the years, that procedure will become fairly easy to do.

We have started to make a case that a newspaper's influence is both measurable, albeit indirectly, and worth measuring. Now let's try to triangulate our target by coming at the issue from a totally different direction. Let's look at the civic journalism movement.

Civic journalism, or public journalism as it was originally called, was denounced by critics as a ploy by publishers to make more money. (One unintended effect of the historic separation of news and business sides has been to give some news people the odd notion that making money is bad.) The concept was introduced into newspaper companies from the top down. In a business so conservative that anything that seems new can set off alarm bells, top-down innovation can create a problem. Nevertheless, it was a really good idea. One of the seminal moments in the movement was a meeting at Key Biscayne, Florida, in the late 1980s, where John Gardner, the founder of Common Cause, told Knight Ridder editorial writers that a news medium's fate and that of its community are strongly connected. Making one strong will strengthen the other. While the word "influence" wasn't used, the concepts were basically the same as in Hal Jurgensmeyer's model from 1978. Civic journalism was a way to use a newspaper's influence to build a stronger polity, and it benefited both the community and the paper. Its most ambitious applications at the outset were in election coverage.

The 1996 presidential election represented a year of great interest in the civic journalism movement. Charitable foundations had taken up the cause, and some put up research money to see if civic journalism could be defined, measured, and linked to measurable effects on the political units that they served. Deborah Potter and I took on this task for the Poynter Institute for Media Studies, the midcareer school for journalists founded by Nelson Poynter in 1975 and enriched by the bequest of his media properties, *Congressional Quarterly* and the *St. Petersburg Times*. We started with a sample of twenty markets, half chosen because we knew that newspapers and broadcasters there were earnestly trying to connect with their communities through civic journalism and the other half because we suspected that their media, with equal earnestness, were sticking to conventional means of election coverage and avoiding the tools of civic journalism.

The idea was to find evidence of the good things that civic journalists wanted to achieve. These included better political knowledge, higher-quality opinions, more participation, and greater trust in media. Those attributes were measured with a pre-test in August and a post-test in late November after the election.[11] But first we had to get a handle on who was practicing civic journalism and to what degree.

Just asking news people if they were practicing civic journalism wasn't quite enough. So many different things were being done in the name of civic or public journalism that there wouldn't be any way of knowing what a "yes" answer meant. So we listed some behaviors that had been proposed as ways to get voters more interested and involved in the election and its coverage. We asked editors and political reporters to tell us the extent to which they planned to do each of the following:

1. Sponsor one or more public forums on issues.
2. Use polls to establish the issues your coverage will focus on.
3. Conduct focus groups with voters to establish their concerns.
4. Form citizen panels to consult at different stages of the campaign.
5. Seek questions from readers and viewers for use when interviewing candidates.
6. Base reporting largely on issues developed through citizen contact.
7. Provide information to help citizens become involved in the political process in ways other than voting.

Their responses confirmed our selection of newspapers, but they did not separate into two well-differentiated groups as we had anticipated. Considering the polarized nature of the debate, we had expected it to be as clearly defined as smoking and nonsmoking sections. On a scale where 300 meant a newspaper was definitely going to do all seven of those things and 100 meant it wouldn't do any of them, our test papers

11. Details of the study are in Paul J. Lavrakas and Michael W. Traugott, eds., *Election Polls, the News Media, and Democracy* (New York: Chatham House, 2000), 113–41.

were spread out all the way between. *The Charlotte Observer* was a 300 and *The Grand Rapids Press* was a 100. But the difference between the bottom of the top 10 and the top of the bottom 10 was quite narrow. *The Columbia State* in Columbia, South Carolina, was at the bottom of the civic journalism group with a score of 237, while the *Chicago Sun-Times* topped the noncivic-journalism category with 231. It turned out that quite a few journalists were using the tools of civic journalism without attaching the novel name.

That was good for statistical purposes because more sensitive tests are available for variables that can be measured on a continuum. To verify that the measured intentions were carried out, we analyzed the content of the newspapers in the period before the election and found a strong correlation between the stated intentions of the reporters and editors and what they put in the paper. The most popular civic journalism strategies at the time were to emphasize issue coverage and minimize polls whose only function was to give candidate standings (horse-race polls). An index that combined those two in the content analysis correlated strongly with the editors' previously stated intent.[12]

These good intentions predicted more. The intent to perform civic journalism was correlated with four of the desired post-election outcomes: knowledge about candidate stands on the issues, trust in the media, social capital, and trust in government. But there was a mysterious gap in the chain of causation.

We thought that intent to do civic journalism would lead to changes in content (true) and that these changes in content would be linked to the attitude and knowledge outcomes. They weren't. The hoped-for effects did show up, but they appeared to be direct consequences of intent without being related to content (Figure 4–1).

Finding the X Factor

What happened? Our measurement might have been faulty. Or, we suggested at the time, some mysterious unknown factor, call it "X," existed earlier in time, and it was the source of both the good outcomes in the community *and* the intent to do civic journalism. It took a couple of

12. Pearson's r = .741.

Figure 4-1

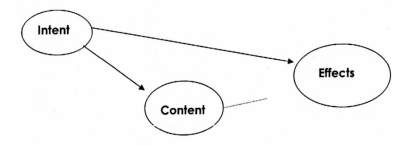

years to figure it out, but eventually David Loomis and I got a handle on what the X factor might be. It might be what we now are calling influence.

Think of it this way. The influence of a newspaper is not created and destroyed with a single issue or even a single election campaign. It is built slowly over the years. It is the result of a healthy relationship between the newspaper and its community—something the civic journalists sought self-consciously to achieve but that would have existed to greater or lesser degree in various places even before the contemporary version of civic journalism was formulated.

How to find out? Conceptually, the problem is easy. First, build a time machine. Then go back a couple of decades and interview the heads of all the newspaper companies you can find to see how much importance they attach to the idea of community and a newspaper's role in making it work. If this is the X factor, then the companies run by those who were community-minded way back in the past will have two things going for them by the time the civic journalism brand name gets tested in the 1990s: (1) citizens who are more knowledgeable and involved and less cynical, and (2) the desire to practice civic journalism in a more formal way.

Rather than wait for somebody to invent time travel, we looked for archaeological traces of civic-mindedness. We found them in the annual reports of all the publicly reporting newspaper companies from 1970 forward. There were nineteen such companies. To make the job manageable, we took those reports from even-numbered years—179 of them—and scanned the message to shareholders from the CEO. Then we ran these messages through an optical character reader and then through the content analysis program DICTION.

What DICTION does is count the frequency of words in a previously designed dictionary. My students and I created two dictionaries, one to define concern for profits and the other to define concern for community. We tested them with live coders who looked at whole sentences and then edited the list down to 137 words for profitability and 102 for social responsibility to create dictionaries that provided a good replication of the human judgments. Here are fifty-word samples from each list:

Profit Dictionary
acquired, advertiser, aggressive, amortization, assets, billion, board, business, capital, cash, centralize, circulation, classified, closed, competitors, complementary, cost, cutting, deal, debt, depreciation, disbursement, diversified, divestitures, dividend, dollars, earn, earnings, economic, economy, efficiencies, efficiency, efficient, expansion, repurchase, restructuring, returns, revenue, rightsizing, sales, shareholders, shares, stock, stockholders, strategic, streamlined, taxes, unprofitable

Social Responsibility Dictionary
awards, celebration, charity, children, citizen, civic, commitment, community, coverage, credibility, culture, donate, editorial, editors, education, ethical, fairness, family, friend, goodwill, honors, honest, honesty, information, integrity, investigative, journalism, local, minorities, mission, Native American, African Americans, news, provider, Pulitzer, quality, readers, reporters, service, social, society, sponsor, stakeholder, student, support, teach, trust, values, volunteer

As you might expect, the words on the profit list were used with much greater frequency than those on the community list. But the ratio varied considerably. To get profit and community on comparable scales, we converted the word counts to standardized scores.[13] Then we subtracted the profit score from the community score. Thus a CEO message with a score of zero would have the same profit-community balance as

13. A standardized score is expressed in standard deviation units. The mean is therefore zero, positive values are above the mean, and negative values are below.

the average across all the 179 messages. A positive score meant a relative tilt toward community, a negative balance meant relatively greater concern for profit. The range of scores was from 4.06, earned by Erwin Potts of McClatchy Newspapers, to -4.51, posted for Arthur O. Sulzberger, Sr., of the New York Times Company. The scores formed a pretty good bell-shaped curve.

Now all we had to do was to figure out which companies were in the civic journalism camp. For an objective evaluation, we turned to Jan Schaffer, who at the time was the director of the Pew Center for Civic Journalism and had a good view of the civic journalism landscape. With her help, we identified eight of the nineteen public companies as having embarked on visible public journalism projects prior to the 1996 national election.

Their scores from the word count were strikingly different. The civic journalism companies scored an average of .590 and the rest were at -.416. That difference spans a full standard deviation, and it is off the scale in terms of statistical significance. If you find this hard to believe, you are not alone. After Loomis and I wrote up our results, we offered the paper to the Association for Education in Journalism and Mass Communication. We were rebuffed with the stern assertion that what we had done was impossible. Only fools would try to predict the behavior of newspaper companies by what their heads had said years earlier, the anonymous judge told us.

But we did. The X factor is real. It is something in the corporate culture of a newspaper company that lets it think about other matters than corporate profit at least part of the time. And that concern for community produced the long-term effects among voters in their communities that we had been originally inclined to attribute to short-term adjustments in content for the 1996 election civic journalism projects.[14]

If we had paid closer attention to the literature of our own field, we would not have been so surprised. There are very few magic bullets in media effects. While public attitudes toward media can blow with the wind, the changes that media make on society come with a slow, drip-drip

14. David Loomis and Philip Meyer, "Opinion without Polls: Finding a Link between Corporate Culture and Public Journalism," *International Journal of Public Opinion Research* 12:3 (Autumn 2000), 276–84.

not quickly or easily detected. And that is going to be a problem for those of us who want investors and managers in news media companies to think about their long-term effects on society. The news business, with its constant emphasis on the latest new thing, finds long-term thinking very difficult. When word got out about my project to investigate the effect of journalistic quality and media credibility on business success, I started getting calls and e-mails from managers who wanted to improve their products in ways that would "move the needle." Because managers are rewarded on the basis of year-to-year performance, this concern with immediately visible progress is understandable. But we should worry that the desire for instant gratification might stand in the way of getting the real job done.

It is very important to focus on trust-raising strategies that reflect a newspaper's character over time. Editors who avoided newsworthy issues because they were controversial might sidestep a temporary downswing in superficial trust at the cost of hurting the inner core of long-term trust.

What are the best trust-raising strategies? The most promising place to look is in the area of basic news-editorial quality.

Leo Bogart looked at the issue in 1977 by asking editors.[15] In one portion of his survey, Bogart asked editors to rank seven newspaper characteristics that he considered subjective, that is, difficult to determine and hard to measure. The rankings:

1. Accuracy
2. Impartiality in reporting
3. Investigative enterprise
4. Specialized staff skills
5. Individuality of character
6. Civic-mindedness
7. Literary style

Editors of large papers were less inclined to give weight to civic-mindedness than those with circulation less than 250,000. Bogart was

15. Leo Bogart, *Press and Public: Who Reads What, When, Where and Why in American Newspapers,* 2nd ed. (Mahway, N.J.: Lawrence Erlbaum), 1989. Bogart drew his list of editors from membership rolls of the American Society of Newspaper Editors (ASNE) and Associated Press Managing Editors (APME). We used only ASNE, whose members tend to be higher-ranking editors, although there is some overlap in membership.

aware of academic efforts to measure newspaper accuracy by asking sources mentioned in news stories but was worried about the effort needed. In the next two chapters, I'll report the results of major attempts to overcome barriers to tracking two of the hard-to-measure qualities—accuracy, and a special subset of literary style that I call "ease of use."

As Esther Thorson's good literature review has shown, most of the studies that relate quality to circulation success are correlation studies. They show that higher quality and higher circulation are linked, but not which is the primary cause. What we probably have, of course, is a reinforcing loop, where quality produces business success which enables more quality. Teasing out the dominant direction of causation is not always possible. A credible effort takes many data points over a much longer period of time than has been spent on any given study. But at least we are reassured that there is connection.

Defining Circulation Robustness

In order to have a consistent measure of business success for this volume, I have settled on circulation because it is audited and publicly available. Raw circulation, of course, is fraught with ambiguity. There are cases like Des Moines (chapter 1) where reducing circulation led to business improvement. My "robustness" measure attempts to bypass that problem, so let me review it again, this time in more detail. It applies three strategies to create apples-to-apples comparisons of circulation:

1. *Focusing on home county circulation.* Most newspapers want to maximize circulation there. It is harder to make a business case for reducing circulation close to the local advertisers than for pulling back from the other end of the state. Retail trade zones tend to be compact and contiguous. Another advantage of using the county is that much administrative data is collected at that level by states and the U.S. Census as well as the Audit Bureau of Circulations. So we get convenience as well as compactness and contiguity.

2. *Dealing with penetration, i.e. circulation divided by households, rather than raw circulation.* High penetration solidifies the relationship of a newspaper with its community, and the advertisers benefit as well. It's

also efficient because a carrier doesn't have to pass so many nonsub-scribing households to deliver to the paying customers. (I noticed this when I delivered *The Clay Center Dispatch* on a bicycle.) And when we use penetration, we are automatically controlling for the direct effect of market size (although there may be indirect size-related effects that we shall also want to consider).

3. *Treating circulation as a dynamic variable by looking at its change over time.* Many long-term pressures are pushing circulation down, and what we really want to know is how effective editorial quality might be as a tool for resisting (or even reversing) that downward pressure. This is the derivation of the concept of "robustness." A robust newspaper is one that minimizes its loss in home county household penetration over a relevant time period. A technical advantage of this strategy is that each newspaper's history acts as its control. Every newspaper, every market is different, and the apples-to-oranges comparison problem is huge. But when we compare each newspaper to itself over time, we are closer to dealing with apples-to-apples.

These strategies will still leave us without firm proof of the direction of causation. Perhaps robustness in penetration is caused by local fac-tors that have nothing to do with content, such as a vigorous circula-tion department. Maybe a successful publisher just likes to cut the editor some slack. Perhaps the best we can do is point to some promis-ing areas for future research.

Here's one. Just as Gaziano and McGrath broke new ground in devel-oping a measure for credibility, two specialists in marketing at North-western University have built a new, improved measure of readership. Bobby J. Calder and Edward C. Malthouse call it the Reader Behavior Score, or RBS. It recognizes that there is more to readership than just looking into the paper.[16]

Like Gaziano and McGrath, they used factor analysis to detect the presence of a latent variable that they conceptualized as representing each person's "total usage of the newspaper." The questions capture four dimensions:

16. Bobby J. Calder and Edward C. Malthouse, "The Behavioral Score Approach to Dependent Variables," *Journal of Consumer Psychology* 13:4 (2003), 387–94.

1. Weekday time spent with the newspaper, categorized in fifteen-minute increments.

2. Weekend time with the Sunday or weekend edition plus time during the week, categorized in half-hour increments.

3. Frequency: respondent checks boxes representing days of the week that he or she reads or looks into the newspaper in an average week. There's a separate box for indicating not reading at all.

4. Completion: respondent is asked how much of the newspaper he or she reads or looks into on an average weekday and an average weekend. The choices are fractions in one-quarter steps with a top category of all or almost all.

The factor analysis verified that the items all measure the same underlying dimension and that they can form a scale with each item given equal weight. Moreover, Calder and Malthouse found that results could be calculated for readership of specific newspapers or for the aggregate of all newspapers read by a respondent (TRBS). This measure has at least two advantages over the "read yesterday" measure that is a cherished tradition in the industry.

One is that it is a continuous measure rather than a binary (yes or no) classification, which means that analysis based on correlation techniques can be used. This kind of analysis is more powerful because it uses more information than the simple categories. The other is that the use of multiple questions reduces error—something that any good reporter knows. It also captures more aspects of the concept, and this ought to make it interesting to all concerned—advertisers, editors, circulation managers. Yet the newspaper industry's immediate response was less than enthusiastic. "Read yesterday" has carved a well-worn groove in newspaper culture and can't be easily replaced or even supplemented.

For an example of where this improved measure might take us, Calder and Malthouse related it to credibility in a very large sample of 37,036 consumers in 101 newspaper markets. Their measure of credibility was less comprehensive than Gaziano and McGrath's but it did include eight items. They asked whether the respondent considered the newspaper accurate, honest, intelligent, experienced, informed about the world and nation, informed about the local community, and whether it represented knowledge and understanding.

Because they had data both at the newspaper level and the individual consumer level, the Northwestern researchers could use a newer statistical technique that not only tests for the readership/credibility connection but also makes it possible to investigate why it is stronger in some markets than others.[17] Their conclusion (as of this writing still unpublished) is that credible newspapers get more readers, and the effect is strongest where competition makes the newspaper fight for its readers. In other words, credibility not only helps, it helps most where it is needed most.

The Calder and Malthouse work was part of the Readership Institute, a project of Northwestern's Media Management Center, an organization that might have the funding and staying power to replicate their massive study enough times to clarify the causation issue. Are newspapers believed because they are read, or are they read because they are believed?

Yes, of course, it's probably some of both. But to convince investors and managers that credibility is good business, we need more precise evidence to estimate how much readership a given increment of credibility can buy and how long an investment in credibility takes to pay out. To that end, the Northwestern project has made a credible start that we can all cheer.

17. The tool is multilevel modeling. It lets the analyst seek out effects on individuals and communities at the same time.

5

Accuracy in Reporting

W H E N *New York Times* reporter Jayson Blair was caught creating fabrications for major stories and putting them in the paper, observers wondered how he got away with it for so long. Why didn't his sources blow the whistle?

It turns out that not complaining is normal behavior for sources who see the truth they've spoken twisted into factual errors in the paper. In the two-year survey of more than five thousand sources conducted for this chapter, only 10 percent of those who reported finding errors said they contacted the newspaper about it. Inexperienced news sources were about half as likely to complain as government officials and others with an ongoing stake in newspaper accuracy.

"I'm too busy," said one. "Not important enough," said another. "I didn't feel it would make a difference," said a third. "They do not take criticism very well," offered yet another.

Not even Donna Leinwand of *USA TODAY* complained when errors in the *Times* led her editors to question her stories on the serial sniper case in the Washington, D.C., area. Blair kept reporting stuff she didn't have, and the editors would ask her to check it out, but the possibility of outright fabrication didn't dawn on her immediately.

"So I would re-call all my sources, enlist others in the newsroom, and spend what turned out to be wasted time trying to track down stuff that Blair had reported. . . . I would reassure my editors, who ultimately do trust me . . . That's not to say I didn't live in extreme anxiety. I would frequently . . . wake up before the newspaper hit my door to see what Jayson Blair had that I didn't. Some of my sources did hint to me that the *Times* stuff was off, but it was hard to tell if they were deliberately trying to mislead to control the investigation. So I'm not sure what I would have complained about while it was ongoing."[1]

In 1988, I asked Lyle Schwilling, an Akron public relations executive and old college friend, what to do when *The Wall Street Journal* misspelled my name. He advised me not to complain. The damage was already done, and a complaint would cause resentment, he reasoned. I was ready to agree until I remembered the electronic database, and I contacted the *Journal,* asking it to fix the spelling there. I received a friendly reply, but, fifteen years later, the correction had not been made.[2]

How can sources and newspapers treat facts so casually when truth-telling is the basic value of journalism? The admonition "Seek truth and report it" is found in the first section of the 1996 revision of the code of ethics of the Society of Professional Journalists. The code goes on to make the advice explicit: "Test the accuracy of information from all sources and exercise care to avoid inadvertent error. Deliberate distortion is never permissible."

It might be argued that this goes without saying. Without a commitment to truth-telling, a news medium's quest for influence would be hopeless. Yet even in the twenty-first century, professional journalists were having trouble creating procedures and checks to minimize error from both carelessness and malice. If it can happen at *The New York Times,* it can happen anywhere.[3]

Newspaper editors in the United States have been sufficiently con-

1. Donna Leinwand, e-mail, October 7, 2003.

2. Gregory Stricharchuk, "Computer Records Become Powerful Tool for Investigative Reporters and Editors," *Wall Street Journal,* February 3, 1988. I was still Philip "Myer" in the database on September 11, 2003.

3. Lori Robertson, "A Choice for Troubled Times," *American Journalism Review* 25:6 (August/September 2003), 9.

cerned about their credibility to commission large-scale studies of the issue since 1985. The most recent, by Chris Urban in 1999, put public concerns about accuracy as the number-one finding. "There is remarkable unanimity between the public and journalists on the fundamental value of accuracy and 'telling it like it is,' but both groups are skeptical about overall accuracy and would rather see journalists get it right than get it first," Urban wrote. "Both journalists and the public believe that even seemingly small errors feed public skepticism about a newspaper's credibility. More than a third of the public—35 percent—see spelling or grammar mistakes in their newspaper more than once a week, and 21 percent see them almost daily."[4]

Estimating Error Rates

The public's estimate of the incidence of errors is quite low compared to that of a group in a better position to know, the sources cited in newspapers. Their perception of error has been of interest to academic researchers since 1936 when Mitchell Charnley of the University of Minnesota got the idea of writing to sources named in the *Minnesota Daily* and inviting them to point out any factual errors. Charnley reported an error rate of .77 per story, or about three errors for every four stories. Others have refined the technique over the years. In a 1980 review of the literature, Michael Singletary reported, "about half of all straight news stories contain some type of error."[5]

Comparison across time and among publications is difficult because the procedure for measuring source-perceived error is not well standardized. Newspapers sometimes do their own version of Charnley's method as a management tool, which includes having conversations with reporters accused of error. With this procedure, it is not possible to promise confidentiality to survey respondents. Without confidentiality, error reporting tends to be understated because sources don't want to risk offending the people who put their names in the paper.[6] With

4. Christine Urban, "Examining Our Credibility: Perspectives of the Public and the Press," American Society of Newspaper Editors (1999), 11.

5. Michael Singletary, "Accuracy in News Reporting: A Review of the Research," *ANPA News Research Reports* 25 (January 25, 1980), 1–8.

6. This source was dramatically demonstrated by Gilbert Cranberg when he replicated a *Des Moines Register* survey with university stationery and a faculty member's sig-

confidentiality, there is a risk of overstating the error rate because sources sometimes define disagreement as error when it is only about tone or emphasis. A real-life example: a source complained that a news story wrongly reported that a contractor was "fired" when, according to the source, the person was actually "removed for defaulting."

In the 1980s, hoping to develop a more rigorous method of measuring error, I worked with *The Charlotte Observer* to create a procedure that provided feedback from reporters whenever they were accused of factual inaccuracy. When reporter and source disagreed on whether an error had occurred, a third party (an assistant managing editor) was asked to judge. This procedure cut the measured error rate in half.[7]

In the search for quality indicators in the current project, a simpler version of the accuracy audit was needed. With the help of Scott Maier of the University of Oregon and some graduate students at Chapel Hill, I ran a pilot project in a familiar venue, the Raleigh-Durham metropolitan area, with three hundred cases for each of two newspapers. After studying the results, we refined the instrument and tried again in the Akron-Cleveland market. Satisfied, we increased the sample specification from three hundred to four hundred and contracted with the Oregon Survey Research Laboratory of Eugene, Oregon, and FGI of Chapel Hill for eighteen more surveys to build, over a two-year period (2002–2003), an accuracy database of more than five thousand news sources named in twenty-two newspapers in seventeen metropolitan markets.

The advantage of this large-scale project is that whatever biases exist in the survey instrument are at least held constant. Cross-market comparisons can be made on the basis of questions that were asked in all of the markets and in the same manner. Questions that were more susceptible to biased response because they dealt with subjective forms of error, for example, "quotes distorted or out of context," were analyzed separately.

We began clipping each newspaper at an arbitrary starting date and taking all of the local stories with unduplicated sources (so that no source would be asked to judge more than a single story) until four hundred stories were clipped. To avoid contaminating the experiment by making reporters aware of it, we delayed mailing until enough news-

---

nature on the covering letter. The *Register* found 14 percent of stories had errors, compared to 63 percent in the academic replication.

7. For details, see Philip Meyer, "A Workable Measure of Auditing Accuracy in Newspapers," *Newspaper Research Journal* 10:1 (Fall 1988), 39–51.

papers to supply the four hundred stories had been collected. (In Detroit and Philadelphia, where two newspapers operate in one market, under a joint agency and joint ownership respectively, the morning and evening papers were combined into a single four-hundred-person sample.)

The initial mailing included a cover letter explaining the survey, a copy of the clipping in which the respondent's name had appeared, and a four-page, self-administered questionnaire with a stamped reply envelope. A reminder postcard went out to everyone seven days after the first mailing. After three weeks, a second, more urgent, letter was sent along with a duplicate questionnaire. When three more weeks had gone by, a final letter and questionnaire were sent. The raw response rate, based on the original four hundred stories and with no adjustment for bad addresses, exceeded 60 percent in every case.[8]

Error surveys tend to get more or less error depending on the number of specific kinds of errors asked about. This survey started with two long lists, one of objective errors such as names, dates, and places, and another of more subjective concerns such as "the story was sensationalized." The range of responses offered to each was from 0 (no error) to 7 (major error). This use of a continuous scale instead of simple yes or no answers makes it possible to do some interesting statistical procedures, including factor analysis to tell us whether the initial division into objective versus subjective kinds of error was valid. It was.

Factor analysis highlights clusters of survey questions that tend to stimulate similar responses. From its results, it became clear that we were dealing with the two basic kinds of error that we had initially envisioned plus a third. Math errors, whether subjective or objective, tended to fall into a class by themselves.

Errors of the first kind deal with categories of information about which usually there is little argument. I call them objective or "hard" errors. The hard errors were defined by responses to the following survey items. The numbers show the frequency with each occurred in the total sample:

My name was misspelled	4%
Other spelling or typographical error	10

8. The Oregon Survey Research Laboratory also used e-mail follow-ups and an opportunity to respond online.

Job title wrong	9
Address wrong	2
Age wrong	2
Location of event wrong	3
Time of event wrong	2
Date of event wrong	2

These seven items were used to form a scale of objective or hard errors. A source who spotted one was likely to find others in the list, a welcome sign of consistency.[9]

Across the total sample (except for Raleigh and Durham where the "other spelling" question was not asked), 21 percent of all stories had at least one error reported in this category.

Mathematical errors were considered in a separate category. There were only two questions, originally designed to represent the soft and hard categories. But respondents didn't really treat them as separate measures. The two items overlapped so strongly that it made sense to combine them into a single scale:

Numbers wrong	13%
Numbers misleading or misinterpreted	13

Among all the newspapers in the sample (again, leaving out Raleigh and Durham), 18 percent of all stories had at least one source-perceived mathematical error.[10]

Turning to the subjective or "soft" indicators of error, we have a list of eleven items. The last three were originally intended to be among the hard indicators, but they correlated more strongly with the soft measures and so are included here:

My quotes distorted or out of context	21%
Interviews with others distorted or out of context	14
I was not identified the way I wanted to be identified	11
Gratuitous references to my race or appearance	2

9. The statistical test for this sort of internal consistency is Chronbach's Alpha. A score of .6 is considered adequate for exploratory research, .8 for confirmation. For these items, Alpha = .754.

10. For the two math items, Alpha = .8.

Story was exaggerated	12
Story was sensationalized	18
Story was understated	10
Other (subjective error)	4
The headline was inaccurate	15
Quotes wrong	21
Other (factual error)	24

Fifty-three percent of all 5,136 stories had at least one of the above source-perceived soft errors.

Put the three categories of error—hard, soft, and math—together, and we find sources claiming that 59 percent of all the stories had at least one error. That's three stories out of five. In other words, if this sample is representative, pick any local story from a daily newspaper, and the chances are better than even that a source mentioned in that story will find something wrong with it.

How important is that? Before comparing the individual newspapers in this study, let's think for a moment about the theory underlying this quest for error. A good theory would tell us what level of error is ideal. I don't know what that would be, but I don't think it's zero.

A newspaper with a zero level of factual errors is a newspaper that is missing deadlines, taking too few risks, or both. The public, despite the alarms raised in ASNE studies, does not expect newspapers to be perfect. Neither do most of the sources quoted in the paper. The problem is finding the right balance between speed and accuracy, between being comprehensive and being merely interesting.

In judging the accuracy of individual newspapers, we can evaluate them in two ways. We can use a normative standard and see which papers are above or below the average for the group as a whole. Being below average in error rate is probably a good thing.

The other way to judge is to find out whether it matters. Does the public think any less of papers with higher error rates? Do error rates affect circulation performance? What about the sources for the stories? When they find errors, are they cynical or accepting?

The remainder of this chapter will try to tackle each of these questions. But first, let's rank the newspapers on the three dimensions of error. Here are the rankings based on the hard errors. The numbers are percentages of stories with at least one error of the indicated type. Please remember that these are approximations based on small samples. The most accurate papers are listed first.

Objective Errors

Rank	Newspaper (cases)	Percent
1	Palm Beach Post (260)	16.2
2	Lexington Herald (280)	16.6
3	Charlotte Observer (315)	16.9
4	Detroit Free Press (133)	18.2
5	Grand Forks Herald (286)	18.9
6	St. Paul Pioneer Press (261)	19.5
7	Aberdeen American News (324)	19.5
8	Tallahassee Democrat (256)	20.3
9	San Jose Mercury News (259)	20.8
10	Detroit News (131)	20.9
11	South Florida Sun-Sentinel (234)	20.9
12	Cleveland Plain Dealer (194)	21.1
13	Wichita Eagle (279)	22.2
14	Duluth News Tribune (272)	22.5
15	Philadelphia Inquirer (135)	22.7
16	Boulder Daily Camera (330)	23.3
17	Akron Beacon Journal (193)	24.4
18	Philadelphia Daily News (129)	26.2
19	Columbus Ledger (253)	27.3
20	Miami Herald (Broward edition) (218)	27.6

Because the samples for individual newspapers are small, the differences between newspapers close in rank are not significant. (The percents are carried to one decimal to break ties in the rank order, not to imply that the estimates are accurate to that level.) Differences between newspapers near the top of the list and papers near the bottom, i.e. where there is at least a six-point difference, are generally significant.[11] Sample sizes are the same for all three error lists.

11. Example: for the difference between the *Detroit Free Press* and the *Akron Beacon Journal*, $F = 13.5$, $p < .02$ (t-test using the 8-interval index).

Math Errors

1	*Detroit Free Press*	12.9
2	*St. Paul Pioneer Press*	13.3
3	*Charlotte Observer*	13.5
4	*Grand Forks Herald*	14.0
5	*Aberdeen American News*	16.0
6	*Wichita Eagle*	16.8
7	*Palm Beach Post*	17.0
8	*South Florida Sun-Sentinel*	17.1
9	*Lexington Herald*	18.3
10	*Broward Herald*	18.4
11	*Duluth News Tribune*	18.5
12	*Columbus Ledger*	18.6
13	*Tallahassee Democrat*	19.1
14	*Akron Beacon Journal*	19.7
15	*Boulder Daily Camera*	20.2
16	*Detroit News*	20.9
17	*Philadelphia Inquirer*	21.1
18	*San Jose Mercury News*	22.0
19	*Cleveland Plain Dealer*	22.2
20	*Philadelphia Daily News*	22.2

The lower incidence of math errors does not mean that errors of this type are less important. It could be just an artifact of the lower number of math-related questions asked in the survey. There is some evidence, as we shall see shortly, that source credibility is especially sensitive to math errors.

It's hard to know what to make of the final class of errors, those that leave more room for some subjective judgment on the part of the news source who may be expressing disappointment over tone or emphasis rather than outright error. We'll tackle that question, but first here's the ranking. This one includes the Raleigh and Durham pilot study because all of the questions in this category were asked in those markets.

Soft Errors

1	*Raleigh News & Observer*	44.0
2	*Aberdeen American News*	46.6
3	*Durham Herald*	47.3

4	*Charlotte Observer*	48.4
5	*Columbus Ledger*	50.2
6	*Lexington Herald*	50.5
7	*Miami Herald* (Broward edition)	51.2
8	*Detroit Free Press*	51.9
9	*Akron Beacon Journal*	52.3
10	*Detroit News*	53.1
11	*Palm Beach Post*	53.8
12	*Philadelphia Inquirer*	54.1
13	*Wichita Eagle*	54.8
14	*Tallahassee Democrat*	55.1
15	*Grand Forks Herald*	55.2
16	*St. Paul Pioneer Press*	55.6
17	*South Florida Sun-Sentinel*	56.0
18	*Duluth News Tribune*	56.1
19	*Cleveland Plain Dealer*	57.7
20	*Philadelphia Daily News*	58.6
21	*Boulder Daily Camera*	60.5
22	*San Jose Mercury News*	62.2

The rankings, in and of themselves, are only mildly interesting. We must next answer two questions. What do they mean? Does it matter? In social science, the validity of a measurement is tested by finding out its predictive power. Learning what other variables of interest correlate with the thing we're trying to measure will give us clues to the answers to both what it means and whether it matters.

We'll start with an internal question. The newspaper industry was convinced by the Urban study in 1999 that accuracy has something to do with a newspaper's credibility. So let's find out if the sources' perceptions of accuracy are related to their perceptions of credibility.

An excellent series of questions to measure credibility was developed by Chris McGrath and Cecilie Gaziano for their ASNE credibility study in 1985. We used all seventeen in the survey of newspaper sources in addition to the accuracy questions. Their credibility items proved to be highly intercorrelated. However, a newspaper editor who wants to replicate this study for his or her own paper needn't go to that much trouble. The six items that are the most representative of the group would do as well. Here are the scores for those items on the twenty-two-newspaper survey.

These are the six items and their mean scores on a scale of 1 to 7 where 1 is the least trusting response and 7 indicates the most trust. (These scales have been reversed from the original questionnaire so that high credibility gets the high numbers.)

Fair—Unfair	5.5
Factual—Opinionated	5.1
Biased—Unbiased	5.1
Can be trusted—Can't be trusted	5.0
Separates fact from opinion—Mixes facts and opinions	4.9
Sensationalizes—Doesn't sensationalize	4.7

Put them together in an index, and the individual newspapers show a wide range of scores. The most trusted by sources is the *Aberdeen American News* at 5.4 on the 1–7 scale. The least trusted is *The Philadelphia Daily News*, a full point lower at 4.4. Of course, we don't know whether these differences are due to characteristics of the newspapers or of their communities.

We can tell, however, that sources who see errors in the paper are less trusting. Dividing our big sample into just two groups, those who reported errors in the stories we sent them and those who didn't, we get:

	Trust level
Sources who found errors	4.9
All other sources	5.6

A more exact way of estimating the effects of perceived inaccuracies on the sources' judgments of the newspapers is to look for correlations. Here's how each kind of error predicts a source's overall trust of the newspaper:

Error	*Variance explained*[12]
Objective error	1%
Math error	5
Subjective error	18

12. The respective correlation coefficients are -.117, -.230, and -.421. All are statistically significant ($p < .001$).

In other words, 18 percent of the variation in trust is accounted for by a source's perception of subjective error. The other kinds of error are significant in the statistical sense, but the size of the effect is trivial.

Why are we not surprised? Respondents to surveys tend to be consistent. A questionnaire that stimulates a memory of a newspaper error is also likely to stimulate a rating of low credibility. The connection between error and credibility would be more impressive if the two measures—error and credibility—came from different groups of people. So let's go outside the accuracy survey and measure newspaper credibility at the community level, using the Knight Foundation's series of community surveys.

The Knight Foundation, as we have seen in earlier chapters, measured credibility with a single question in telephone surveys of the adult population in 1999 and 2002. The responses are volatile, and so, as in earlier chapters, we'll use the mean of those two measurements for an estimate of the long-term credibility of the newspapers in the Knight Foundation communities.

These communities, at least the ones that we are using here, are identified by whole counties. The Knight credibility question asks about the newspaper with which the respondent is most familiar, which is no problem where one local newspaper dominates a county. In counties with two strong papers, we need to combine them by taking the mean of their separate measures of source credibility. This is done regardless of ownership: the *Detroit Free Press* and *The Detroit News* in Wayne County, Michigan, are competitive editorially but managed by the Detroit Newspapers, a joint agency of Gannett and Knight Ridder. In Philadelphia, the *Inquirer* and *Daily News* are both owned by Knight Ridder. In Broward County, Florida, the *South Florida Sun-Sentinel* and *The Miami Herald* are separately owned and operated. Each pair is treated as one for this particular analysis.

The result is strong validation for the opinions of sources about the credibility of the newspapers in which they are quoted. These people are not merely venting their personal prejudices and concerns. Their ratings and those of the public track quite closely. This is best illustrated with a scatterplot. As source credibility increases, so does population credibility. For each 1-point increase on the 7-point source credibility scale, the percent of the population who believe all or most of what they read in the paper increases by 13.3 points.[13]

Figure 5-1

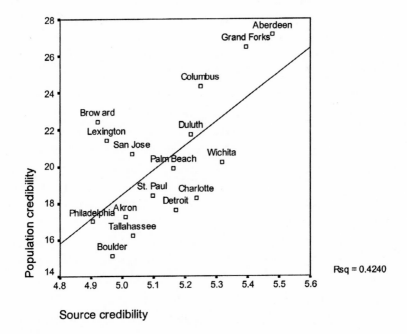

Source credibility

Source credibility is measured here as the average score across all seventeen credibility items in the accuracy survey. The scale items have been reversed from the original questionnaire so that 7 is the most trusting response. Population credibility is the percent of adults interviewed by telephone who report that they believe the paper with which they are most familiar all or almost all of the time.

This gives us confidence that the opinions of sources about the credibility of the newspaper are worth attending to. And it makes us wonder what helps or undermines that credibility. When we looked at individual-level data (5,136 cases), we saw that perception of any kind of error undermines credibility, but the subjective error category

13. Correlation = .651, p = .006. With the separately owned and managed papers in Broward County removed, r = .757, p = .001, slope = 16.1.

is the worst. Moving to market level after several newspapers have been combined or dropped, we have only sixteen cases, but that relationship still holds.[14] One reason might be that sources have more emotional reactions to errors in the subjective category, and so they give them higher marks on the 1–7 scale (minor to major). If we ignore those marks and just count the frequency with which each kind of error occurs without regard to the intensity, math errors move to the front.[15] In other words, among sources, minor math errors can cause as much distrust as major soft errors. This suggests rational judgment on the part of sources. Math errors are not ambiguous, and so it only takes a small one to trigger distrust. Subjective errors are ambiguous, and sources recognize this and discount them to some extent—but not enough to keep them from being an important source of lost credibility.

Those communities whose newspapers have greater than average errors have less trusting citizens. This effect is mostly indirect, mediated through the sources themselves. Figure 5.2 shows the path diagram with the partial correlations.

At the market level (sixteen cases), when source credibility is controlled, the correlation between error and population credibility vanishes. When error frequency is controlled, the effect of source credibility on population credibility remains strong. This is pretty good evidence that sources have a greater effect on the population than the direct effect of the errors by itself.

The Two-Step Flow

Paul Lazarsfeld and his colleagues found something similar when they studied the 1940 presidential election in Erie County, Ohio. They discovered that the direct effects of media on the population were less powerful than personal contacts. Ideas flow, they said, from media "*to* the opinion leaders and *from* them to the less active sections of the population*" (emphasis in original).[16]

Newspapers with a high density of errors have the least trusting

14. Correlation = .620, p = .01.

15. At the market level, number of stories with any math errors correlates with source credibility at $r = -.602$, $p = .005$.

16. Paul Lazarsfeld, Bernard Berelson, and Hazel Gaudet, *The People's Choice* (New York: Duell, Sloan and Pearce, 1944), 151.

Figure 5-2

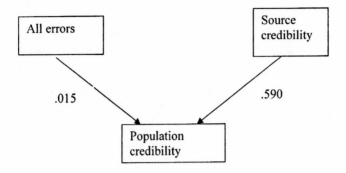

sources, and their skepticism filters down to the population. If this re-sult seems strange, remember that the sources tend to be the elites. Ordinary people do get interviewed, but not as often as opinion leaders.

The different kinds of error yielded different kinds of effects when we switched from individual-level analysis to community-level. Many things can cause that, but one obvious possibility is that influence is not dis-tributed evenly among the individuals in either the community survey or the accuracy survey. The bottom line is the effect on credibility in the community. Here the results point in the same direction. When credi-bility as measured by the Knight Foundation surveys is the criterion, errors make a difference. This finding tends to vindicate Urban and other credibility investigators who have pointed to factual errors as a source of credibility problems.

Now we can get very serious. Is accuracy, or its lack, a business prob-lem?

As in previous chapters, the test will be home county penetration ro-bustness. Because the accuracy data were collected in 2002–2003, we'll use a comparable basis for the robustness measure, the Audit Bureau of Circulations County Penetration Reports for 2000 and 2003. The reports were released in the spring of those years and are based on the most re-cent audit in each case. Because the audit periods vary, I have (as in earlier chapters) adjusted for their different lengths by calculating the months between audits and producing an annualized robustness figure.

It is easy to interpret. A robustness value of 1.0 means that a news-paper maintained level home county penetration in the average twelve-month period between the audits. A value of .9 means it has kept only 90 percent of its previous penetration in an average year. Most of the values are less than 1.0 because newspaper circulation has been declining, while the number of households has been growing.[17] The home county test might not be ideal for every newspaper because some have important strategic interests outside the home county. But if a newspaper can't do well in the place where it is created, it's probably not doing well at all, and the county definition has the virtue of being uniform—which, of course, is why the Knight Foundation settled on it for its community studies.

For this research question—does accuracy affect robustness?—we shift to yet another unit of analysis. This time it is the individual newspaper. We can do this because we have accuracy data and circulation data for twenty-two specific newspapers—just what we need to see if accuracy matters.

And it does matter. As before, the most powerful predictor is the source's assessment of the newspaper's credibility.

The effect is clearer when we adjust for market size. Larger markets have a harder time holding on to home county penetration than smaller ones as a general rule. Detroit is an exception because in the 2000–2003 period it was coming off the devastating effects of a strike against both papers that lasted eighteen months and ended in February 1997. After 2000, the newspapers started to recover some of that loss.

The adjustment makes only a small difference, but it leads to significant correlation between source credibility and robustness.[18] An increase of one point on the source-perceived credibility scale (running from 1 to 7) is associated with seven additional percentage points of penetration robustness. Newspapers that are trusted by the people mentioned in their columns are more successful.

Let's stop and think about this for a minute. Does this result mean that a successful newspaper is one that panders to its sources and does anything to keep them happy? Of course not. A good newspaper cannot keep all of its sources happy all of the time. Truth is impartial. It helps some and hurts others, and a newspaper cannot yield control to its sources. The old adage still holds: "Tell the truth and let the chips fall where they may."

But it would be wrong to conclude that source-perceived credibility

17. Penetration is defined as circulation divided by households.
18. Correlation = .422, p = .05.

is always based on self-serving reporting by the sources. If their answers rewarded servile behavior by reporters and punished truth telling, there would not be such a high and significant correlation between the sources' evaluation of newspaper credibility and the community's evaluation as measured in the Knight Foundation surveys. One of the most important results of this accuracy study is that sources are pretty good judges of a newspaper's quality.

And why shouldn't they be? Their names are in the paper, and they will naturally give it a closer reading and more careful judgment than the average reader can or wants to manage. Moreover, many, if not most, news sources are opinion leaders. They form attitudes toward the paper based on how its writers handle matters on which the sources are informed. You can bet that they don't keep those opinions to themselves. The two-step flow theory of communication has been well documented since Lazarsfeld originally noticed it.

But let's get down to something more direct. Elite opinion about the paper is a fuzzy measure compared to the count of different kinds of perceived error. Let's turn to the firmer measures of accuracy—soft error, hard error, and math error. All are positively associated with penetration retention in the period closest to the accuracy study: 2000–2003. All hint at or lean toward statistical significance individually, and they do meet the traditional test collectively. In other words, the fact that all three point in the direction predicted by the theory that accuracy matters is significant. To give you an idea of the magnitude of each measure's importance, here is a listing by the amount of variance in robustness that is explained by the error rate.

Type of error	*Variance explained*[19]
Math error	12%
Hard error	8
Soft error	6

Because the three kinds of error are correlated with one another, the effects are not cumulative. You could throw them all into one multiple-

19. Statistical significance for the three correlations is .11, .21, and .29 respectively. Their combined probability, however, is .007. A one-tailed test, which is justified because theory points us in only one direction, would produce an even smaller figure.

regression equation, and you would still get only fractionally more than 13 percent of variance explained. The explanation is obvious: a newspaper that is careless with one kind of fact is likely to be careless with all.

Here's another way to look at it. For each newspaper, add up the three error scores for an index of total error. For each additional percentage point of any kind of error, penetration robustness in the period 2000–2003 declines by a tenth of a point per year. It's a small effect, but accumulating drip by drip over time, it is devastating.

The amount of variance explained by the totality of source-perceived reporting error is small enough to remind us that many other factors are at work in explaining the long-term decline in circulation penetration. The main importance of accuracy is as a building block in creating long-term credibility for the newspaper.

The case for caring about accuracy—and for caring about how the persons mentioned in the news feel about accuracy—is now strong enough that we are justified in wondering about the sources of error and the relative pain that different kinds of error cause. This survey can shed some light on both.

The scale in the accuracy survey allowed each respondent to rate each error on a seven-point scale from "minor" to "major." Errors in the subjective category got the worst ratings, especially judgments that the story was exaggerated (mean rating of 3.40) or sensationalized (3.35). Inaccurate headlines were also considered relatively serious (3.15). Misspelled names and other spelling errors were not considered major (1.90) unless the source's own name was misspelled (3.04). Getting the date of an event wrong was considered more serious (3.13) than getting the time wrong (2.86). Simple numerical errors were viewed more tolerantly (2.95) than interpreting the numbers in a misleading or erroneous way (3.14).

The top reason given by sources, when asked to judge why the reporter made a mistake, was simply that the reporter didn't understand what he or she was writing about. Among those who noted errors, more than one source out of four (29 percent) had this complaint.

The complete list of attributed causes, in order of their frequency:

Reporter didn't fully understand the story	29%
Pressure to get the story done on time	23

Not enough research	16
Events surrounding the story were very confusing	15
Laziness on the part of the news staff	12
Reporter didn't ask enough questions	12
Reporter didn't ask the right questions	12
Pressure to get the story before other media	8
I gave the reporter wrong information	1

Those top three reasons form a logical cluster. All are related to a newspaper's capacity, a topic to be discussed more fully in a later chapter. Think back for a moment to the SPJ code and its advice for reporters: "Test the accuracy of information from all sources and exercise care to avoid inadvertent error." It means not taking the first source you talk to at face value and spending the time to come at the facts from more than one angle.

A brief case study will make this clearer. When University of Florida researchers released their 2000 statewide Florida Health Insurance Study, a newspaper reporter asked the head of his local health care district about it. The local official said the report was flawed because it underestimated the county's population by more than 20 percent.

But it hadn't. A closer reading would have revealed that the survey was not based on the total population of the state, but only on the non-Medicare population, persons under the age of sixty-five. Checking with a second source who knew more about it would have helped. The reporter did indicate an attempt to reach a state official who wasn't available.

In this situation, a typical reporter could claim that no error was made. He talked to a source who was in a position of authority, and he accurately reported what that source said. He also correctly reported a failed attempt to get another view. How much responsibility does a reporter have to go behind what officials say? Under the passive reporter model of the 1950s, none at all. What officials say is news. After Senator Joseph McCarthy, Republican from Wisconsin, used that passivity to scare the public about Communists in the government, reporters started to get more aggressive. By today's standards, the Florida story was in error. And all three of the top reasons in the list were in play here: the reporter didn't fully understand the story, there was pressure to get it in the paper before another source could be reached, and the needed research didn't get done.

A newspaper that is understaffed will be more susceptible to this kind of problem than one that is not. How much difference does staff size make? If you look at the ratio of news-editorial staff per thousand circulation, the answer is not much. The correlation is small and insignificant. But it turns out that you can explain a lot of error just by looking at the extent to which the size of the news-ed staff is related to home county circulation. This a rather weird way to measure capacity because it favors newspapers that concentrate their effort locally over those that have far-flung news interests. But when we do it, the number of staff members per one thousand home county circulation explains 26 percent of the variation in total error.[20]

It also explains 36 percent of the variation in number of stories with math errors.[21] These estimates are based on just those seventeen cases where I was able to combine staff information from the American Society of Newspaper editors with circulation data and the accuracy survey. (There would have been nineteen, but two large newspapers had staff ratios that were wildly out of line with the rest, a consequence of their aggressive coverage of surrounding counties. I left them out.) As part of the deal with ASNE, I can't reveal the names of the papers in this analysis, but I can show you the scatterplot. Here's the one for math errors.

As the relative staff size increases, the rate of math error drops. The regression line shows the overall rate of the drop. An additional staff member per one thousand circulation in the home county is good for a 4.4 point decline in the error rate. That's an average or mathematical expectancy, of course, because the data points (representing newspapers) are somewhat scattered around the straight line representing the 4.4 point rate of drop. The scatter happens because factors other than sheer staff size also affect error rates.

Like what? How about the ratio of copy editors to reporters? It makes a small difference, but leaning toward statistical significance, in the case of mathematical errors. For each additional reporter whose work a

20. Correlation = .515, p = .034.
21. Correlation = .598, p = .011.

Figure 5-3

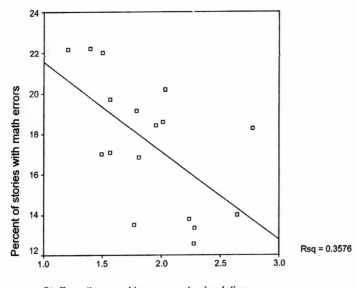

Staff per thousand home county circulation

copy editor has to watch, the math error rate goes up two percentage points.[22]

The business case for accuracy seems clear. Let's summarize it with a picture:

Accuracy improves robustness and population credibility with only small direct effects. The main effect is through source credibility. The resulting boost to population credibility improves the robustness of circulation. With a larger sample and the opportunity to study accuracy over time, we could make the case even stronger. If we had such a record we could tell managers and investors what specific danger points to watch for, and we could equip them with a distant, early warning of a future business failure. For now, all we can say for sure is that accuracy is good.

There is an obvious related issue not measured in my study. In addition to the number of reporters and copy editors, the competence of those people is bound to make a difference. Journalists need a working

22. Correlation = .395, p =.077.

Figure 5-4

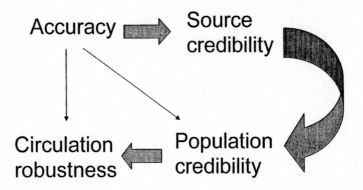

knowledge of the subjects that they cover, and they need programs for lifelong learning. If the newspaper industry could see the value in mid-career training, it might pay for it. Instead, it relies largely on charitable foundations.

The other contribution of this chapter is that it provides a baseline by which other newspapers can compare their performance. For editors and managers considering their own accuracy studies, the 2002–2003 questionnaire is reproduced here.

Figure 5-5

NEWS ACCURACY SURVEY

——— ——— ———
mail ID code

To assess and improve news accuracy and credibility, this questionnaire seeks your comments on a news story in which you provided information or commentary. Please read the story and answer the following questions. You and your organization will not be identified in any published results.

1. Sometimes news stories contain information that is wrong, such as names, addresses, and titles. The following is a list of different types of factual errors that can occur in a news story.

    For each item on the list, please circle 0 if the error did not occur. If it did occur, circle the number that corresponds to the severity of the error--with 1 standing for a minor error and 7 for a major error.

Type of Error			**Severity of Error**					
	None	**Minor**						**Major**
The headline was inaccurate	0	1	2	3	4	5	6	7
Numbers wrong	0	1	2	3	4	5	6	7
My name was misspelled	0	1	2	3	4	5	6	7
Other spelling or typographical errors	0	1	2	3	4	5	6	7
Job title wrong	0	1	2	3	4	5	6	7
Address wrong	0	1	2	3	4	5	6	7
Age wrong	0	1	2	3	4	5	6	7
Location of the event wrong	0	1	2	3	4	5	6	7
Time of the event wrong	0	1	2	3	4	5	6	7
Date of the event wrong	0	1	2	3	4	5	6	7
Quotes wrong	0	1	2	3	4	5	6	7
Other factual error	0	1	2	3	4	5	6	7

2. If errors were made, why do you think they happened? From the list below, *please place a check* mark beside every reason for errors that you think may apply to the story we sent you.

    ❏ Laziness on the part of the news staff

    ❏ Pressure to get the story done on time

    ❏ Reporter didn't fully understand the story

    ❏ Events surrounding the story were very confusing

    ❏ Pressure to get the story before other media got it

    ❏ Not enough research

    ❏ Reporter didn't ask the right questions

    ❏ Reporter didn't ask enough questions

    ❏ I gave the reporter wrong information

    ❏ Other (please specify)_____

Please turn the page

3. Sometimes news stories contain information that is technically correct but misleading, or that give the wrong impression. Following is a list of different ways that a news story can mislead.

    For each item, please circle 0 if the problem did not occur. If it did occur, circle the number that corresponds to the severity of the error--with 1 standing for a minor error and 7 for a major error.

Type of Error				Severity of Error				
	None	Minor						Major
Essential information was omitted	0	1	2	3	4	5	6	7
My quotes distorted or out of context	0	1	2	3	4	5	6	7
Interviews with others distorted or out of context	0	1	2	3	4	5	6	7
I was not identified the way I want to be identified	0	1	2	3	4	5	6	7
Numbers in the story were misleading or misinterpreted	0	1	2	3	4	5	6	7
There were gratuitous references to my race or appearance	0	1	2	3	4	5	6	7
The story was exaggerated	0	1	2	3	4	5	6	7
The story was sensationalized	0	1	2	3	4	5	6	7
The story was understated	0	1	2	3	4	5	6	7
Other (Please specify)_____	0	1	2	3	4	5	6	7

4. Did you contact the newspaper to correct the error or to point it out?

    ☐   Yes
    ☐   No (Why not?)_____

5. Please rate the *overall* story according to the following criteria.

	None						Major error(s)
Seriousness of errors combined	1	2	3	4	5	6	7
	Fair						Unfair
Fairness and balance	1	2	3	4	5	6	7
	In Context						Out of context
Context and perspective	1	2	3	4	5	6	7
	Clear						Confusing
Clarity of the story	1	2	3	4	5	6	7
	Appropriate						Inappropriate
Emphasis given to the story	1	2	3	4	5	6	7
	Newsworthy						Frivolous
Newsworthiness	1	2	3	4	5	6	7
	Factual						Opinionated
Story tone	1	2	3	4	5	6	7
	Information						Sell papers
Story motivation	1	2	3	4	5	6	7

Please turn the page

6. **Please rate the credibility of each of the following news sources:**

	Credible					Not credible	
Local television news	1	2	3	4	5	6	7
National network news	1	2	3	4	5	6	7
Local radio	1	2	3	4	5	6	7
National Public Radio	1	2	3	4	5	6	7
The local newspaper you read most often	1	2	3	4	5	6	7
The national newspaper you read most often	1	2	3	4	5	6	7
Your favorite Internet news source	1	2	3	4	5	6	7
News magazines	1	2	3	4	5	6	7

7. **In the context of this news story, which category best describes your role:**

- ❑ Government official
- ❑ Business representative
- ❑ Citizen activist
- ❑ Witness/bystander
- ❑ Commentator/expert
- ❑ Community member
- ❑ Other (please specify)_____

8. **Have you been interviewed for a newspaper story before?**

❑ No ❑ 1-3 times ❑ 4-9 times ❑ 10 or more times

9. **Are you:**

- ❑ Male
- ❑ Female

10. **What year were you born?**_____

11. **How much formal education do you have?**

- ❑ Some High School or Less
- ❑ High School Diploma
- ❑ Some College
- ❑ College Degree
- ❑ Some Graduate Study
- ❑ Graduate Degree

12. **Based on your experience with this news story, please rate your willingness to serve as a news source again.**

Eager   **Reluctant**
1   2   3   4   5   6   7

Please turn the page

Here are some pairs of words and phrases with opposite meanings. Please *circle the number* between each pair that best represents how you feel overall about the newspaper that printed the story we sent you.

Fair	1	2	3	4	5	6	7	Unfair
Unbiased	1	2	3	4	5	6	7	Biased
Tells the whole story	1	2	3	4	5	6	7	Doesn't tell the whole story
Accurate	1	2	3	4	5	6	7	Inaccurate
Respects people's privacy	1	2	3	4	5	6	7	Doesn't respect privacy
Cares what readers think	1	2	3	4	5	6	7	Doesn't care what readers think
Watches out for your interests	1	2	3	4	5	6	7	Doesn't watch out for your interests
Concerned about the community's well-being	1	2	3	4	5	6	7	Not concerned about community's well-being
Separates fact from opinion	1	2	3	4	5	6	7	Mixes fact and opinion together
Can be trusted	1	2	3	4	5	6	7	Can not be trusted
Doesn't sensationalize	1	2	3	4	5	6	7	Sensationalizes
Moral	1	2	3	4	5	6	7	Immoral
Patriotic	1	2	3	4	5	6	7	Unpatriotic
Concerned mainly about public interest	1	2	3	4	5	6	7	Concerned mainly about making profits
Factual	1	2	3	4	5	6	7	Opinionated
Reporters are well-trained	1	2	3	4	5	6	7	Reporters are poorly trained
Ethical	1	2	3	4	5	6	7	Unethical

Please fold the questionnaire and return it in the enclosed self-addressed, stamped envelope.

The completed survey should be mailed to:

Professor Scott Maier
Oregon Survey Research Laboratory
5245 University of Oregon
Eugene, Oregon 97403-5245

Please accept the enclosed dollar bill as a token of our thanks for your participation. The clipping is yours to keep.

Thank you for completing the survey

6

Readability

W H E N Blanche Perkins taught journalism at Clay County (Kansas) Community High School in the 1940s, she told her students to "write for a 12-year-old child." Edwin A. Lahey, as the Washington bureau chief for Knight Newspapers in the 1960s, advised his reporters to "write for people who move their lips when they read."

At United Press in midcentury, the maxim was, "Write for the Omaha milkman."[1] In that same period, Associated Press editors said, "Tell it to Sweeney and the Stuyvesants will understand. But tell it to the Stuyvesants and the Sweeneys may *not* understand."[2]

Writing for a broad audience is a constant struggle. Journalists today are well educated and have broad interests, and their natural inclination, if not checked by self-monitoring and good editing, is to write for each other. The marketplace does not discourage this elitist tendency as

1. Reynolds Packard, a former UP foreign correspondent, wrote a novel based on his experiences and called it *The Kansas City Milkman* (New York: Dutton, 1950).

2. Cited by Alan J. Gould in the Foreword to Rudolf Flesch, *The Art of Readable Writing* (New York: Harper and Brothers, 1949).

efficiently as it used to. That is because journalism is in slow transformation from appealing to a mass audience to tailoring its appeal to many different classes of audiences. As Alvin Toffler has said, we are moving from "a few messages sent to many people to many messages sent to a few people."[3]

Newspapers, and the advertisers who support them, still count for the most part on maximizing their reach, so clear writing is still important.

The message was carried most forcefully to the media by a 1938 immigrant from Austria named Rudolf Flesch. His dissertation at Columbia University included a mathematical definition of readability based on sentence and word length, and it was published as *Marks of Readable Style*. But, being a dissertation, it was not itself a very good example of readability. When he realized this, Flesch rewrote it as *The Art of Plain Talk*, published by Harper and Brothers in 1946. His approach caught on quickly, and he became a consultant for the AP, whose head called Flesch's work "one of the most significant developments of our journalistic times." A friendly critic suggested that his advice was more about rewriting than writing, so Flesch thought it over, created a revised formula, and included more detailed advice in *The Art of Readable Writing* three years later.

His formula survives today in a number of software packages, including the most popular word processor, Microsoft Word. Here it is. If you have a little statistical knowledge, you will recognize it as a linear regression equation with two independent variables:

Reading Ease Score = 206.835 − 1.015(SL) − .846(NS)

where SL represents the average number of words per sentence and NS is the number of syllables per 100 words. The formula yields a measure from 0 to 100 where 0 is practically impossible to read and 100 is easy for any literate person.

Flesch validated his formula by testing the ability of persons with different education achievement to read materials of varying reading ease scores. He reported that a score of 70 to 80 made the text accessible to persons who read at the sixth grade level, while scores in the range of 30 to 50 required some college education.

3. At a conference of Knight Ridder editors in Point Clear, Alabama, ca. 1979. Toffler developed this idea in *The Third Wave* (New York: Morrow, 1980), 171–83.

A measure can be valid, however, without being reliable. Validity is the degree to which a measure can be verified by external criteria—the tests of comprehension by actual readers in this case. Reliability describes the extent to which repeated measures with different judges will produce the same answer. Ironically, the use of computers has complicated matters. They make it easy to analyze large volumes of text, but without much assurance that two different computer programs—or even two different releases of the same computer program—will produce the same answer with the same formula.

That's because of a lack of agreement on the best algorithms for measuring sentence length and counting syllables. You can estimate the number of syllables by the number of vowels. For example, the word "vowels" has two of each. But you need to find a way to deal with double vowels as in "deal" and "double." Sometimes a terminal vowel will denote a syllable, as in "syllable" and sometimes not, as in "denote."

Sentence length is more straightforward. When the computer encounters a space, it can assume a word has ended, and a period can mean the end of a sentence. But how does the computer distinguish a decimal from a period or a sentence-ending period from one that identifies an abbreviation, as in "Dr. Flesch"?

And how does it deal with spaces in a numbered or bulleted list? Since there is no obvious answer, different programmers have come up with different answers.

The folks at Microsoft like to tweak the algorithm for the Flesch formula in their popular word processor, and so it changes from one edition to the next. Consider the following passage:

There are two arguments against this plan: 1. It is too expensive. 2. It is impractical.

The Flesch engine in Word 2000 returns a Reading Ease score of 72. But if you use Word 2002, you get a score of 50.6 for the very same text. Between the two releases, the Microsoft code writers decided that numbered lists with periods lowered reading difficulty.[4] That might well be an improvement, but it's something to watch out for if you are making comparisons of measures taken at different times and with the same software issued in different versions.

4. Reported at http://support.microsoft.com in Microsoft Knowledge Base Article 267964 (retrieved April 29, 2003).

For this chapter, I have relied on a program called Readability Calculations supplied by Micro Power and Light Co. of Dallas, whose spokesperson reports that "we have used the same formula algorithms . . . for many years."[5] Testing it on a convenience sample of twenty articles from the *Greensboro* (N.C.) *News & Record,* I found that it produced scores that were moderately correlated (r = .747) with those from Word 2000 but averaged 1.4 grade levels lower. It therefore provides a more generous estimate of reading ease than Word 2000.

Flesch's success inspired imitators, and I have experimented with seven different readability tests that can be performed by a computer. In my newspaper sample, they are highly intercorrelated, which is good evidence that they all measure the same concept. The most popular alternative to Flesch is Robert Gunning's Fog Index. It uses the average number of words per sentence plus the percentage of words that Gunning defines as "hard." The hard words are all those with three or more syllables with the exception of proper names, compounds of small words, and words that would be only two syllables if not for ending in "ed" or "es." In my sample of 2,125 newspaper stories, Fog and Flesch-Kincaid are almost perfectly correlated.[6]

So I stick to Flesch because it is easy to understand and can be expressed as either a school grade level or on the 0–100 scale.

And that raises another issue. If we are going to use readability as one dimension of newspaper quality, what shall we define as "good"?

The Grade-Level Standard

When Dr. Flesch tested readers in the 1940s, grade levels in school had different meanings than they do today. As high school education has become nearly universal and college education much more common, the grade levels for given achievement have tended to drift upward. As a college teacher, I hear my older colleagues complain, "We're teaching high school now." It seems that way because college freshmen read less than we did when young, and they arrive with a different kind of literacy, one more attuned to pictures and less to words. Flesch's word test needs to be renormed.

---

5. Ed Frantz, personal communication, April 30, 2003.
6. r = .990.

Dr. Peter Kincaid, a psychologist out of Ohio State University, did just that for the U.S. Navy in 1974. The navy was worried that its training manuals were too difficult for Vietnam-era recruits to read, and it wanted to create readability specifications for the contractors who wrote them. Kincaid tested 569 sailors at the Naval Air Station at Memphis and Great Lakes Naval Training Center at Chicago. Then he renormed the Flesch test, and created the formula that uses the same sentence and word information but yields a score that represents a different grade level.[7]

In the thirty years between Flesch and Kincaid, real educational attainment appears to have slipped by one or two grade levels. Flesch estimated that a Reading Ease score of 70 to 80—which he called "fairly easy"—would be about right for a person with a sixth-grade education.[8] Applying Kincaid's renormed grade level formula to my newspaper sample and comparing it to the original Reading Ease scores, I find that the average story in that same 70–80 range of reading ease is classified by the Kincaid formula as eighth-grade material. Similarly, Flesch's "standard level," which is reading ease of 60 to 70, was estimated by him as requiring a seventh- or eighth-grade education. After the Kincaid renorming, it calls for between a ninth- and tenth-grade level.

As I write, the 1974 norm is still the most recent. Kincaid followed Flesch by three decades, and now another three have passed, and somebody should renorm the test again. But let's not wait. For the rest of this chapter, I'll express readability in terms of the Flesch-Kincaid grade level as measured by the algorithm in Micro Power and Light Co. Readability Calculations Release 3.7e.

To help you get an intuitive feel for the Flesch-Kincaid scores, here are some examples:

"The Mast Cell in Inflammatory Arthritis," by D. E. Woolley in the *New England Journal of Medicine* for April 23, 2003, contains phrases such as, "The inflammatory processes that result in rheumatoid arthritis are multifactorial, involving complex interactions among the cytokine network . . ." It has a Flesch-Kincaid grade level of 13.9.

7. Telephone interview with Peter Kincaid, principal scientist of the Institute for Simulation and Training, Orlando, Florida, April 28, 2003. Dr. Kincaid is also affiliated with Central Florida University.
8. Flesch, *The Art of Readable Writing,* 149.

John F. Kennedy's inaugural address, ". . . ask not what your country can do for you . . ." scores at 10.3.

William Faulkner's Nobel Prize acceptance speech, in which he said, "I decline to accept the end of man . . ." was written at level 8.8.

Patrick Henry's "give me liberty or give me death" speech tests at 6.6.

Analysis of articles in forty newspapers published in 2003 suggests that the typical staff-written newspaper story would fall in the middle of the above list, written for a reader with a ninth-grade (by 1974 norms, remember) education. Is that good or bad? Since the median adult (age twenty-five and older) in the 2000 census had an educational attainment a bit beyond twelfth grade, it might appear that newspapers are succeeding in casting a fairly wide net. But they are not.

As Flesch pointed out in *The Art of Readable Writing,* most people like to read below their ability level. It's easier. Besides, reading takes practice, and some of us tend to lose the habit. "The *typical* reader for each readability level will usually be found in the next higher educational bracket—and sometimes the stretch will be even wider," said Flesch (emphasis in original). "Time magazine, which boasts of its readership among top executives, is written in breezy, high-school English; and Presidents and Supreme Court Justices have been known to devour extremely easy-to-read mystery stories."[9]

All of this points to an ideal Flesch-Kincaid target for newspapers somewhere in the sixth- to eighth-grade range, not so far from Blanche Perkins's advice to her high school students. But there is one more complication.

To use readability as an indicator of quality for a given newspaper, we ought to know its business plan. Readability is not a one-size-fits-all measure. A paper that is aiming at an affluent subset of the population can justifiably be harder to read than one that seeks to sweep in every literate person. Later in this chapter, we'll try to assess the extent to which newspapers are aiming upscale and abandoning the goal of speaking to an entire community defined by geographic or political boundaries.

But first, let's take a look at the lay of the land. The sample population of newspapers for this study comes from three sources:

9. Ibid., 150.

1. Twenty-nine newspapers in the communities followed by the Knight Foundation and for which survey data are available.

2. Eight newspapers in counties where social indicators were collected at the county level by Robert Putnam's Saguaro Seminar.

3. Three newspapers added because they provide interesting examples of competition within their metropolitan areas.

Please remember that this is a convenience sample and not directly generalizable to the entire population of daily newspapers in the United States. All we can say is that the list is interesting and diverse, although it includes a disproportionate number of newspapers owned by one company, Knight Ridder.

My graduate assistants first collected a sample from each newspaper by downloading all of the local stories from the online version of the newspaper on each of six different weekdays spread across a six-week period starting in January 2003. This constructed-week method avoids concentrating the sample in one narrow time period. We did some spot-checking to see if the online and print versions of the stories were the same. They were. In most cases, the stories were collected on the day of publication or the day after. In the few cases where online access was not available, we prevailed upon newspaper librarians to send us the hard copies.

This procedure yielded a grand total of 2,125 stories from forty newspapers. Only one story in four was readable at the eighth-grade level or lower. Here is the quartile breakdown:

Most readable fourth:	Grades 4–8.1
Second most readable:	Grades 8.2–9.5.
Third most readable:	Grades 9.6–10.7
Least readable fourth:	Grades 10.8–13.9

The case is clear. Many newspaper stories are too hard to read.

To compare newspapers, I aggregated the scores by newspaper with the reading grade level average weighted by story length. The ranking, from most to least readable, looks like this:

Table 6-1

Rank	PAPER	Average grade level	Number of stories
1	Grand Forks Herald	5.04	59
2	Centre Daily Times	7.80	50
3	Myrtle Beach Sun News	8.22	63
4	Columbia The State	8.34	58
5	Philadelphia Daily News	8.68	45
6	Biloxi Sun Herald	8.74	31
7	Arizona Republic	8.78	53
8	Charlotte Observer	8.88	60
9	Aberdeen American News	8.89	27
10	St. Paul Pioneer Press	9.03	57
11	Akron Beacon Journal	9.03	51
12	Palm Beach Post	9.08	60
13	Detroit Free Press	9.10	56
14	Long Beach Press Telegram	9.13	47
15	Wichita Eagle	9.14	55
16	Cleveland Plain Dealer	9.27	59
17	Columbus Ledger-Enquirer	9.35	47
18	Durham Herald-Sun	9.43	56
19	Detroit News	9.45	60
20	Kalamazoo Gazette	9.47	59
21	Fort Wayne Journal Gazette	9.52	58
22	Philadelphia Inquirer	9.53	58
23	Macon Telegraph	9.53	55
24	Duluth News Tribune	9.54	40
25	Tallahassee Democrat	9.55	37
26	Syracuse Post Standard	9.60	51
27	Greensboro News and Record	9.65	59
28	Bradenton Herald	9.79	47
29	Gary Post Tribune	9.81	64
30	Boca Raton News	9.84	49
31	Baton Rouge Advocate	9.86	60
32	San Jose Mercury News	9.86	57
33	Raleigh News and Observer	9.86	59

Table 6-1 (*cont.*)

Rank	PAPER	Average grade level	Number of stories
34	Miami Herald (Broward Edition)	9.86	60
35	Fort Wayne News Sentinel	9.88	54
36	Miami Herald (Home Edition)	10.00	60
37	Lexington Herald Leader	10.04	44
38	South Florida Sun Sentinel	10.04	46
39	Boulder Daily Camera	10.10	55
40	Houston Chronicle	10.38	59

To get a better picture of what we are talking about, let's look at an excerpt from one of those fifth-grade-level stories from the *Grand Forks Herald:*

A few combines were rolling in Walsh County fields this week, but it is tough going, said Craig Askim, Walsh County extension agent.

"They're not getting it all; they're just getting the spots they can," he said Tuesday. "I saw combines stuck in the field yesterday."

Across the Red River in Minnesota, 78 percent of the spring wheat was harvested as of Sunday, compared with 93 percent last year and the five-year average of 86 percent, according to the Minnesota Agricultural Statistics Service.

In Polk County, Minn., at least a couple of farmers have about a third of their spring wheat left to harvest, said Jochum Wiersma, University of Minnesota small grains specialist. More than 3 inches of rain fell in some areas around Crookston last week, leaving soils saturated.

"We have standing water in a lot of fields. We're now in trouble. It's not even that the grain has to dry off, it's that it is going to be tough harvesting. Some people apparently tried over the weekend where they had less rain, and they were making big ruts trying to get in the fields."[10]

10. Ann Bailey, "Agriculture: It's Always Something," *Grand Forks Herald,* September 4, 2002.

You see where the good score comes from. There is a mix of short
sentences and long ones, and there are few words of more than two syllables.

Now, for contrast, let's go to an excerpt from a story written at the
twelfth-grade level for the *Houston Chronicle.*

> The chronically congested freeway now has one reversible high-
> occupancy vehicle lane, three regular lanes in each direction and two
> two-lane frontage roads, making 11 through lanes in all, plus en-
> trance and exit lanes at intersections.
>
> The planned widening from the West Loop to Texas 6 calls for two
> toll lanes in each direction, plus four or five through lanes and three
> frontage lanes, making 18 or 20 through lanes plus the entrances and
> exits.
>
> From Texas 6 to the county line at the city of Katy, the freeway
> would be widened to four lanes in each direction, plus three frontage
> lanes. The present diamond lanes would remain for buses and multi-
> occupant vehicles, said Transportation Department spokeswoman
> Janelle Gbur.
>
> The idea of using toll lanes to speed the widening project was in-
> tended to plug any gaps in federal highway funding that might cause
> delays, cutting construction time from 10 years to six.
>
> The Harris County Toll Road Authority will provide $250 million
> of the expansion cost, raising money through the sale of bonds,
> which will be paid off from revenues from the toll road system. The
> Federal Highway Administration will pay for the bulk of the remain-
> der of the project.[11]

The sentences tend to run on. The writers could have said "cars with
passengers" instead of "multi-occupant vehicles." Or "often choked" instead of "chronically congested."

These two examples provide a sense of the range between newspaper
stories in the top and bottom quarters on the readability scale. Does any
of this matter? We have several things to check, including household
penetration and, in the case of the Knight Foundation communities,
credibility. But first, let's refine the measure a little bit by adjusting it for
the education level in each community.

11. Rad Sallee and Steve Brewer, "Metro Wins Space for Rail on Katy," *Houston Chronicle,* September 4, 2002.

This is fairly easy to do because we are using the newspaper's home county as the unit of analysis. The 2000 census gives us the educational attainment of adults (age twenty-five and over) by county. Although the census reports in categories instead of mean grade level, we can take the midpoint of the categories and estimate a mean grade level of educational achievement. The range is from 10.37 (a first-semester high-school junior) in Biloxi, Mississippi, to 12.77 in Boulder, Colorado (some college).

The Readability Stretch

The next step is to subtract that mean achievement from the newspaper's Flesch-Kincaid score. The result, negative in every case, indicates how far below the average citizen's grade level the newspaper is stretching in order to encompass a broad readership.

College towns have an advantage in this scoring system because of their high overall educational attainment. The *Grand Forks* (North Dakota) *Herald* and the *State College* (Pennsylvania) *Centre Daily Times* top the list.

So here is another long table. Remember that a high negative value for "readability stretch" is good. It indicates an effort to reach a large proportion of the potential audience in the home county.

You will notice that we have lost two newspapers here. The Long Beach *Press Telegram* is a medium-size paper in a huge county, Los Angeles. The *Boca Raton News* is a tiny paper in a medium-size county, Palm Beach. Each has less than 4 percent home county penetration, which makes county-level statistics for these papers poor measures. It is best to drop them for this part of the analysis.

Now we can attack a question that bothered us earlier in the chapter. If writing for Sweeney instead of the Stuyvesants really works, the papers whose editors are pushing the writing down the grade-level scale ought to be enjoying higher penetration. And they are. The correlation is far from perfect, but it does achieve statistical significance. The reading stretch explains more than 16 percent of the variance in home county penetration.[12]

Here's a simpler way to compare. Let's divide the thirty-eight newspapers into two groups, high and low on the readability stretch measure.

12. $r = .406$, $p = .011$.

Table 6-2

Stretch rank	PAPER	Grade level rank	Readability stretch
1	Grand Forks Herald	1	-5.77
2	Centre Daily Times	2	-4.84
3	Columbia The State	4	-3.06
4	Tallahassee Democrat	24	-2.80
5	Boulder Daily Camera	37	-2.67
6	Charlotte Observer	8	-2.66
7	Durham Herald-Sun	17	-2.65
8	St. Paul Pioneer Press	10	-2.53
9	Myrtle Beach Sun News	3	-2.46
10	Raleigh News and Observer	31	-2.31
11	Philadelphia Daily News	5	-2.30
12	Syracuse Post Standard	25	-2.29
13	Akron Beacon Journal	11	-2.22
14	Palm Beach Post	12	-2.08
15	Kalamazoo Gazette	19	-2.05
16	Aberdeen American News	9	-1.99
17	Cleveland Plain Dealer	15	-1.96
18	San Jose Mercury News	30	-1.89
19	Arizona Republic	7	-1.81
20	Wichita Eagle	14	-1.76
21	Biloxi Sun Herald	6	-1.63
22	Lexington Herald Leader	35	-1.56
23	Fort Wayne Journal Gazette	20	-1.55
24	Greensboro News and Record	26	-1.51
25	Philadelphia Inquirer	21	-1.45
26	Detroit Free Press	13	-1.37
27	Macon Telegraph	22	-1.33
28	Duluth News Tribune	23	-1.33
29	Columbus Ledger-Enquirer	16	-1.20
30	Miami Herald (Broward Edition)	32	-1.20
31	Fort Wayne News Sentinel	33	-1.19
32	Baton Rouge Advocate	29	-1.15
33	Bradenton Herald	27	-1.04

Table 6-2 (*cont.*)

Stretch rank	PAPER	Grade level rank	Readability stretch
34	South Florida Sun Sentinel	36	-1.02
35	Detroit News	18	-1.02
36	Gary Post Tribune	28	-.77
37	Miami Herald (Home Edition)	34	-.53
38	Houston Chronicle	38	-.15

(The cutting point is -1.8). Those with high stretches had a mean home county penetration of 45 percent. The low-stretch group averaged only 37 percent.

Moving from penetration to raw circulation, what should we expect? Well, larger papers have more resources, so they should have more copy editors browbeating reporters into writing clearly. High circulation should be associated with higher readability.

It's not. If anything, there's a bit of a tendency for larger papers to write more densely. There is a fairly obvious theory to fit this. Anyone who has worked the spectrum from small-town to big-city newspapering knows that small-town reporters are closer to their communities, and those in larger markets are in more danger of being socially isolated. They could be the ones most anxious to write for one another rather than for the general public. So up goes the grade level.

OK, let's try something a little more sophisticated. Maybe papers with a high editing capacity have better readability. The copy editors at such papers would have more time to consider such issues of quality control. There are several ways to measure editing capacity: news-ed staffers per one thousand circulation, ratio of reporters to copy editors, ratio of reporters to editors and managers. I tried them all.

None of it has any measurable effect on readability. It's as though readability is something that is ingrained in the culture at some newspapers and not at others. Capacity is bound to matter on many important things, but not, it appears, this one.

What's more, ease of use might not even be the result of a coherent management strategy. If it were, there should be a correlation between readability and other ease-of-use factors such as proportion of stories with photos or other illustrations. When my students tested a subset of thirty-two newspapers for this correlation, they found none.

Now that brings us to a really interesting question. Think about the newspapers toward the bottom of the readability stretch scale, let's say the bottom quarter, from *The Sun-Herald* in Biloxi, Mississippi, to the *Houston Chronicle*. Are they that way because their managers are aiming for a more upscale audience than the rest? Are they engaging in editorial redlining?[13]

If you were to eavesdrop on the pitches that newspapers make to advertisers, you might think so. They emphasize the education and income of newspaper readers as opposed to nonreaders. You might almost think they were glad not to have the nonreaders. Or it might just be a rationalization to cover up a readership decline that is beyond their control.

These data give us a way to tell. All we have to do is compare the readability stretch to advertising rates. If newspapers are using reading difficulty to shut out poorer, less educated people, and if that strategy works, the papers that are hardest to read relative to audience education should charge the most (per thousand circulation) for advertising. Their ads would be worth more because the papers skim the cream of the buying power in the county.

They are not. If we look at published rates, there is a tiny effect, but it does not even approach statistical significance. And it is in the opposite direction from what was expected. Papers that have a high readability stretch to get a broad audience ask for about five cents more per one thousand circulation for a Standard Advertising Unit.

Those are published rates. We can get a better picture by looking at actual advertising rates. Here we are limited to the twenty-one Knight Ridder newspapers in the sample because, thanks to its helpful management, we have the mean rates for 2002.[14]

The correlation is positive and significant.[15] The more a newspaper stretches its readability to reach for a broad audience, the more it charges

13. The expression "redlining" comes from a twentieth-century practice by real estate dealers of promoting residential segregation by marking specific neighborhoods as limited to one race or another.

14. My thanks to Steve Rossi, president of Knight Ridder's newspaper division, for supplying the data.

15. $r = .466$, $p = .03$.

per thousand for its advertising. This is a direct refutation of the redlining theory.

If we look at the scatterplot, we see that the significant effect is driven by two small college towns in relatively isolated locations. State College, Pennsylvania, and Grand Forks, North Dakota, both have educated populations, and there is not much competition for print advertising. So we should probably wait for more evidence before declaring that higher readability has a direct effect on a newspaper's advertising. But the indirect effect—more readability, more readership, more circulation—should be enough. And the redlining hypothesis is not supported. The newspapers in this sample might be thinking about shutting out poorer citizens, but they're not doing it.

The next obvious thing to check is credibility. Here we are limited to the twenty-seven counties where the Knight Foundation has measured newspaper credibility. The comparison cannot be direct, because the readability measure is specific to individual newspapers, while credibility is measured across all locally circulated newspapers, whatever their origin, in each county. And we find no correlation, no evidence that counties with easy-to-read newspapers have people who are more trusting in their papers.

For a bottom-line measure, we have our old friend, circulation robustness, to compare. Are the more readable newspapers more successful at resisting the long-term decline in home-county penetration?

On the whole, no, they are not. But they are not less successful either. There is no correlation between readability or readability stretch and robustness. For this test, I used the spring ABC county penetration reports from 2000–2003. The more readable newspapers, as measured in late summer 2002, were neither more nor less robust in their home county circulation penetration. Perhaps if there were more papers like the *Grand Forks Herald* in the highest readability range, there would be enough variation to detect an effect. Or perhaps other ease-of-use measures would tell us more. A newspaper's effectiveness at storytelling through compelling narrative might make a difference.

For now, all we can say is that being readable doesn't appear to hurt. If a newspaper wants to make its content accessible to a wider range of the citizenry as a public service, there is no visible cost to doing so. In this world, there are very few cost-free ways of helping society work better, and so it ought to be done.

7

Do Editors Matter?

O N C E while attending a newspaper marketing seminar in Durango, Colorado, I took a side
trip to visit the Mesa Verde cliff dwellings. They
were built by a long-vanished people who were
part of the migration from Asia, across the Bering
land bridge, down to South America. This migration, historians believe, was so gradual that
no person who was a part of it realized there
was a migration at all.

The decline of newspaper readership has
had the same ethereal quality. While total
newspaper circulation as a proportion of
households was clearly in decline from
the 1920s, most people at the turn of the
century still read at least one newspaper
on most days. The loss was in readers
of more than one paper, and it was
caused by the weeding out of duplicate circulation in multi-newspaper
markets.

Even with that obvious explanation, the decline in household
penetration was enough of a public relations problem to make
newspaper publishers look for a
different measure. The Newspaper Advertising Bureau, an
arm of the old American
Newspaper Publishers As-
sociation, stepped in to find

a solution. The result was elegant. It helped advertisers focus on readership rather than circulation with the "read yesterday" measure. Developed by Leo Bogart with endorsement from the Advertising Research Foundation, it was a careful and conservative measure that corrected for the tendency of survey respondents to overreport their reading behavior because of its social desirability.

The trick was to ask a series of questions about specific newspapers that the respondent had read or looked into in the previous week. That introductory question vented the social desirability bias. Then, with the fact that the person was a reader established, he or she was asked about each paper claimed: "When was the last time before today that you read or looked into the (name of paper)." If the person volunteered "yesterday," he or she was counted as a reader. By spacing the interviews appropriately through the week, researchers could convert "read yesterday" into a valid and reliable measure of average daily readership.

Bogart's path-breaking survey was released in 1961. It showed that on a typical weekday, 80 percent of adults read a newspaper. This "near-universality of readership" became "the basic theme of every sales presentation delivered to advertisers."[1] The pitch was so effective that "read yesterday" became the gold standard of newspaper readership measurement. The Simmons Market Research Bureau, which did audience measurement for TV and magazines, added the newspaper measure to its annual surveys. But then something unexpected happened. The number began to fall.

For Bogart, the trouble started in 1971 when the bureau's own survey showed read-yesterday readership three points below the most recent Simmons number of 78 percent. Methodological error was suspected. Respondents had not been asked about distant newspapers with local household penetration of less than 5 percent, leaving out people who might, for example, read *The Wall Street Journal* and nothing else. After much review and argument, the bureau cranked in a point-and-a-half correction to represent the national papers. From there, Bogart recalled much later, "we could easily round the number to 77 percent . . . this reduced the discrepancy between our survey and SMRB's 78 percent to a single percentage point—a relatively innocuous difference."

---

1. Bogart, *Preserving the Press,* 36.

The presentation to advertisers, which used to claim newspaper readership by four out of five Americans, was modified to say "nearly four out of five." Yet, recalled Bogart, "while I went about the country enthusiastically putting on this presentation . . . I was uneasily conscious that the percentage was really closer to three out of four than to four out of five."

Not long after that, my then employer, Knight Ridder, started detailing me to an occasional readership study along with my reporting duties in the Washington bureau. When the American Society of Newspaper Editors was in town, a gathering of Knight Ridder editors was organized, and I was asked to tell what the minuscule decline in readership might mean.

"If it happens once," I said, "it could be a sampling fluke and nothing to worry about. If it happens twice, we should start to worry. If it happens a third time, it's an earthquake."

James K. Batten, then editor of *The Charlotte Observer*, recalled that observation years later when he was Knight Ridder vice president for news and on his way to becoming CEO. "It was an earthquake," he said.

The story of the newspaper industry's response has been told in detail by Leo Bogart in his 1991 memoir *Preserving the Press*. The Newspaper Readership Project, the pooled effort of a number of newspaper trade associations, operated from 1977 to 1983 on several fronts. They included advertising and promotion of newspapers, circulation improvement, and audience research.

From an economic point of view, the most effective response was the simplest: compensate for fewer readers by raising prices to advertisers. From 1975 to 1990, publishers pushed advertising rates up by 253 percent, even though newsprint prices were up by only 161 percent and the Consumer Price Index gained 141 percent.[2] Charging more for delivering less is essentially a liquidation and harvesting strategy, but this was not the conscious goal. There is no evidence that the owners had given up on the business and were hoping to cash in and get out.[3] Instead, publishers and investors congratulated one another on newspapers' pricing flexibility. Like the migrants from Asia to South Amer-

2. Ibid., 53.
3. Porter, *Competitive Advantage,* 310.

ica, they couldn't see the movement toward the brink because it was so slow.

The editors of newspapers adopted a more alarmist view. They tended to take the readership decline personally and blamed themselves for not providing sufficiently compelling content. Readership studies became a growth industry for a time. Manipulating content costs little or nothing, and so trying to halt the readership decline by tailoring the editorial product to readers' wants seemed particularly attractive. Leo Bogart suspected otherwise.

Bogart was hoping to find the editorial formula for success when, as I described in chapter 4, he surveyed editors on their definition of a quality newspaper in 1977. Then he compared the responses from editors of successful newspapers with those of editors whose papers were slipping, expecting to find the secret of success. But the winning editors and the losing editors, to his surprise, gave the same answers!

". . . editors of successful and unsuccessful newspapers seemed to be operating by identical editorial philosophies," Bogart recalled later. "The inevitable conclusion seemed to be that the forces that made a newspaper lose circulation were largely independent of its content. Success or failure had more to do with pricing, distribution, and population changes in the cities where papers published than with the character of the editorial mix or the operating practices or theories of individual editors."[4]

The editors were not ready to hear this. Rather than rejoicing that the readership decline was not their fault, they attacked Bogart's survey and his conclusions. Any outcome was better than facing the possibility that they were powerless.

This development had a profound effect on my career. Throughout the 1970s, I had become a roving precision journalist, traveling from my base in the Washington bureau to help papers in the Knight Ridder group apply social science research methods to their local news stories, mostly dealing with race, poverty, and opposition to the war in Vietnam. These efforts were well received, and top management at the company decided that the same methods could be applied to the readership problem. That's how I found myself detailed to the occasional marketing

---

4. Bogart, *Preserving the Press*, 108.

study. By 1976, these assignments were sufficiently frequent that I moved out of the National Press Building to an office in Reston, Virginia, near my home and, more importantly, my mainframe computer supplier. Data were still entered on punched cards, and analysis was done with mainframes in those days. I began the transition from Harvard Data-Text, an excellent higher-level computer language for the obsolescent IBM 7090 series, to SPSS, which ran on a great variety of newer machines.

Two years later, I left reporting altogether and was posted to Miami to become Knight Ridder's first director of news and circulation research. My mission was a parochial version of Bogart's. While he was trying to halt or reverse the readership decline for the newspaper industry in general, I was attempting to do that job for Knight Ridder in particular. A number of entrepreneurial research firms jumped into the fray.

The idea of newspapers doing market research at all was controversial. Some journalists regarded it as pandering to lowbrow reader tastes. One of the entrepreneurs, Ruth Clark, helped fuel the controversy with a series of focus groups undertaken for ASNE that provided support for a softer approach to the news. Information from focus groups, of course, is not generalizable, and her report began with the obligatory disclaimer. As soon as that was out of the way, she started to generalize. Newspapers, she told the editors in 1979, should be "more caring, more warmly human, less anonymous." To compete with television, her argument went, newspapers needed to be more entertaining.[5]

Five years later, she ran a larger study, funded by United Press International, which included survey research that was generalizable, and she reversed her course. Now reader interests had moved from entertainment to news. If she believed that her original focus groups were wrong, she never acknowledged it. Readers, she implied, had changed in that short time.

We were all failures. The readership decline proceeded in straight-line fashion, with only a hint of leveling off in the early 1980s as the baby boomers passed the age of thirty and began acquiring the community ties that are associated with newspaper readership.[6] The Readership Project ran out of funding and shut down. I moved to academe

5. Ibid., 142.
6. The leveling is visible in the General Social Survey. See the chart in chapter 1.

and compiled my work in *The Newspaper Survival Book: An Editor's Guide to Marketing Research.*[7] Bogart summed up his effort in a far more comprehensive volume, *Press and Public: Who Reads What, When, Where and Why in American Newspapers.*[8]

After the Readership Project

That pretty much left it up to the academy to come up with new measures and ideas. Bogart's work did not go unnoticed, and two good researchers at Michigan State University built on his 1977 survey to try to challenge the dismal notion that what editors do doesn't matter.

Bogart's sense that editors were off the hook, you will remember, was based on his survey asking them to define quality in news. The basis of his no-effects conclusion was that editors of successful and not-so-successful newspapers expressed the same news values.

But there are some other things to consider. When Stephen Lacy and Frederic Fico did the study mentioned in chapter 4, they realized that editors' values might not be a sufficient measure of what they actually do. The capacity of individual editors to act will vary. Two kinds of capacity are involved here. One, placed under intense scrutiny by Rick Edmonds for the Poynter Institute, is based on the resources that an editor is provided by his or her publisher and can be measured by size of staff, newsroom budget, library resources, training effort, and the like.[9]

Another element in capacity is neither financial nor physical, but intellectual. If two editors have the same news values as measured by Bogart, and the same human and physical capacity as measured by Edmonds, the cleverer of the two might be able to convert those resources into better and more effective content than his or her less talented peer. I call this the "Bellows effect" after Jim Bellows, the editor who made a career of moving from one failing newspaper to another, propping each one up with ephemeral brilliance.[10]

Lacy and Fico tackled both problems simultaneously with refreshing

7. Philip Meyer, *The Newspaper Survival Book: An Editor's Guide to Marketing Research* (Bloomington: Indiana University Press, 1985).

8. Bogart, *Press and Public.*

9. Rick Edmonds, "Measuring News Capacity: A First Cut at an Indicator," www.poynter.org/content. Posted May 21, 2002 (retrieved June 5, 2003).

10. Jim Bellows, *The Last Editor* (Kansas City, Missouri: Andrews McMeel, 2002).

directness. They assembled a sample of 114 newspapers for their constructed week (seven different days of the week, not sequential, to avoid being overly affected by a single unusual news event). Then they adapted the following measures based on Bogart's editor survey.

> Ratio of staff-written to wire service and feature copy.
> Amount of nonadvertising content (news sections only).
> Ratio of interpretation and backgrounders to spot news reports.
> Ratio of illustrations to text.
> Number of wire services carried.
> Length of stories in news sections.
> High ratio of nonadvertising to advertising content (news sections only).

Three measures considered more important by editors in the survey were given a greater weight than lesser ones.

Fico and Lacy built an additive index from these variables, using standardized scores (meaning that each variable was expressed in terms of its own deviation from the mean). The next thing they needed to proceed with the execution of Bogart's original intent was to find a measure of newspaper success.

They chose circulation as reported by the *Editor and Publisher* Yearbooks of 1985 and 1986. The Lacy-Fico constructed week for sampling quality was in November of 1984. Circulation reports lag the audits by several months to a year, so the presumed cause and effect were close together in time.

Taking circulation as the sign of success requires, of course, a correction for the size of the market, and this creates another problem, defining the market. Newspapers' own definitions of their markets are highly idiosyncratic, tailored to delivery constraints, commuting patterns, and retail trade zones. The researchers in this case chose city population as an indicator of market size, which could be a reasonable surrogate for actual market size in most cases, especially in the smaller markets. (In larger places, the main newspaper sometimes does better outside the central city than it does inside, owing to the deterioration of downtown, a problem that Bogart detected early on.)

A straightforward way of correcting for market size would have been to work with household penetration, which is circulation divided by households, and is easy to interpret. A newspaper with 50 percent penetration in a given area has circulation equal to half the number of households. (Market researchers prefer households to population because one copy of a newspaper is typically shared by all the members of a household, regardless of the number of people there.)

Instead, they used a quite different method of adjustment, using city population and newspaper quality to predict circulation in a regression model. Multiple regression is a convenient way of estimating the effects of different factors on your ultimate object of interest. Lacy and Fico reported that with this procedure, which tended to level the playing field for circulation size, news quality explained about 22 percent of the remaining variation in circulation. In short, better newspapers sell more copies.

That would be wonderful news except for the problem that haunts us throughout this book. There is no very good way to tell whether greater circulation is the cause or the effect of higher quality. Bigger circulation brings in more revenue, leading to economies of scale that can free up more resources for the newsroom. And, of course, causation could run both ways: as a virtuous cycle if higher circulation leads to still more quality and as a vicious cycle if both are slipping. One way to sort this out is by introducing time as a variable.

Even if the quality measure by Lacy and Fico had clearly been made earlier in time than the circulation measure, there would still be the problem that newspaper circulation does not change much from year to year. Slow change means that circulation at Time 1 will always correlate with circulation at Time 2. One needs measures of both quality and circulation at clearly different times—with enough lag to produce an effect—plus a form of analysis that adjusts for the autocorrelation, that is, the tendency of a variable measured at different points in time to correlate with itself.

In their recommendations for further research, Lacy and Fico said that an update of Bogart's 1977 work would be useful. Koang-Hyub Kim and I took them up on it and ran a partial replication in 2003. It was a partial replication because we used only members of ASNE while Bogart also sampled members of the American Press Managing Editors. Many editors belong to both organizations, and ASNE was able to provide

significant logistic help, so we gratefully concentrated on its membership. Using a mixed mail and Internet methodology, we received returns at a rate of just over 50 percent.

In his search for readily applicable yardsticks, Bogart had given editors a list of twenty-three criteria and asked them to rate the importance
of each on a scale of +3 to -3. Then he averaged the scores and ranked
the items. Focusing on the fifteen items that editors had ranked the
highest for Bogart (and adding some of our own), we came up with relative scores that were very close to those obtained fifteen years earlier.
The basic values of editors had not changed.[11]

Among those that were used in both 1977 and 2002, the same top
ten emerged, although in different order. Here they are:

1977	2002	
1.	2.	High ratio of staff-written copy to wire and feature service copy
2.	1.	Total amount of nonadvertising content
3.	8.	High ratio of news interpretations and backgrounders to spot news
4.	5.	Number of letters to the editor per issue
5.	3.	Diversity of political columnists
6.	6.	High readability on Flesch or similar scoring systems
7.	10.	High ratio of illustrations to text
8.	9.	High ratio of nonadvertising content to advertising
9.	7.	High ratio of news to features
10.	4.	Number of staff bylined features

One of the values promoted by the Hutchins Commission in 1947,
to go beyond mere reporting of facts and provide interpretation that
would yield "the truth about the facts," appears to have shifted as the
importance of interpretations and backgrounders fell from third to eighth
place. On the other hand, diversity of political columnists gained some-

11. "Quantifying Newspaper Quality," presented to AEJMC, Kansas City, Missouri,
July 30, 2003.

what in importance, as did number of staff features, and both of these can be construed as bows to the Hutchins recommendation. The values of editors are remarkably stable.

To reduce all the data from our survey to a smaller number of manageable concepts, Kim and I applied factor analysis, the tool developed for psychological testing decades ago. We were seeking clusters of question items that correlate with one another, an indication that all are measuring the same underlying factor. There were five such clusters. In order of importance ascribed to them by editors, they were:

1. Localism
2. Interpretation
3. Editorial vigor
4. Quantity of news
5. Ease of use

Bogart had avoided asking about qualities that he considered too subjective to measure. Accuracy and literary quality were among these. (Accuracy was dealt with in chapter 5 and literary quality, interpreted as readability, was considered in chapter 6. In this analysis, the Flesch scores from chapter 6 will be used as the indicator of ease of use.)

Following the example of Lacy and Fico, I then looked for relationships between each of these five factors and business success. Instead of using circulation as the indicator of success, I chose household penetration in the home county as my static measure and the robustness of that penetration—its staying power over time—as the dynamic measure.

This strategy is fine in theory, but shares the same limitation that Lacy and Fico faced: the content measures come from a single brief period in time, November 1984 in their case and April 2002 in mine. This is fine if we assume that newspaper culture so inhibits change over time that the brief snapshot of content is an adequate indicator of how a paper was edited over a longer period. Their content sample was a full week, including Sunday. I stuck to weekdays. Since I collected circulation penetration data for 1995, 2000, and 2003, a number of possible effects can be sought. But because of the limitations of the content measurement, we should expect some frustration and consider this research more for exploration than confirmation.

Localism

Let's start with localism, measured on a sample of thirty-two newspapers. This is a convenience sample, consisting of papers in communities supported by the Knight Foundation communities plus a few others for which community data was available. The sample consists of four days in April 2002: April 11, 12, 16, and 17.

Coders examined every news story in the front and local sections of each newspaper and classified them according to source: staff-written or wire.[12] A column of briefs was considered a single story and counted fractionally if its individual components came from both kinds of sources. Staff-written stories filed from remote locations were counted as local on the theory that the purpose of sending staff far afield is to obtain reports of remote events that are tailored for the local audience.

The mean across the thirty-two newspapers was a staff-written rate of 46 percent. But the range was wide: from 64 percent at the *Detroit Free Press* to 31 percent at *The Macon Telegraph*. Intuitively, one might think that smaller papers, being more focused on their communities, would have a higher percent local, but this is not the case.

Perhaps they can't afford it. A fairly large market might be required to support heavy local coverage. The evidence is in the way that localism increases steeply with size of market (defined by home county) up to about three hundred thousand to four hundred thousand households. For the markets larger than that, size did not matter. Apparently, critical economies of scale accumulate rapidly at first, then more slowly. They vanish altogether once a market reaches four hundred thousand households. Below that, the ability of a newspaper to staff up for heavy local coverage is quite dependent on market size.[13]

The effect is clearest if we plot it with market size on the horizontal axis and percent local stories on the vertical. The diminishing effect can be visualized as a straight line by reexpressing market size as its logarithm, which has the effect of exaggerating effects at the low end of the

12. My thanks to Owen Covington and David Freeman who performed this work in the summer of 2002.

13. The effect is best illustrated by a log transformation. The correlation between percent local and the log of the number of households is .629 ($p = .0001$). Using the raw number of households, correlation is significant but less impressive: .427 ($p = .015$).

scale and minimizing them at the high end. (Imagine a plot on a rubber sheet. The log scale stretches it out on the left or low side and compresses it on the right or high side.)

Figure 7-1

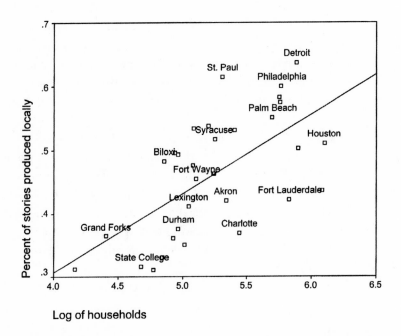

Log of households

Some of the papers are identified by their cities to help you see the effect of market size.

Before we accept the conclusion that the size effect is simply a matter of papers in larger markets having more resources, let's consider another way to look at it. Papers in smaller communities don't need to go to as much effort to maintain good household penetration. The dynamics, and sometimes the isolation, of a smaller community enhances demand for a newspaper.[14] Perhaps smaller communities get by with less local coverage because they don't need it as much to be successful in business.

14. This, too, is an effect that becomes less pronounced as community size increases. The correlation between household penetration and the log of households is -.656, p = .00005.

Whatever the reason, a paradoxical result is presented. As the proportion of local stories in a newspaper increases, household penetration declines. Lest some publisher or investor take this as an excuse to demand thinner local staffing, let me hasten to add that there is no causal relationship here. How do I know that?

We return to the clever statistician's trick used in earlier chapters that lets us take the effect of market size out of the equation. It is partial correlation. When the playing field is leveled in this way, that bothersome negative relationship between household penetration and percent local stories vanishes into the ether. It does not turn positive. It just goes away. Both localism and low penetration are functions of market size. Nothing more.[15]

For another reading, my student coders classified the same set of stories not by who wrote them, but by the location of the news event. The proportion of all stories from all sources that had a state or local origin was used as another indicator of localism. Nothing interesting turned up. With or without a statistical adjustment for size of the market, the use of local geography instead of staff writing to denote localism was unproductive. The geography-defined localism correlated with neither penetration nor robustness.

But localism is just one indicator of quality. Let's keep looking.

Editorial Vigor

Ralph Thrift, Jr., was a graduate student at Oregon when he invented his index of editorial vigor. This is an additive index, as opposed to a scale, because the presence of one of its factors does not imply the presence of the others. But the more of them that are present, the more vigorous is the editorial effort. This is obvious on its face, hence it has what social scientists call "face validity."

In Thrift's conceptualization, the following elements contribute to the vigor of an editorial:

1. Localism. It takes more vigor to discuss local matters that your readers know about.

15. With log of households as control, $r = -.098$, $p = .60$.

2. Controversy. Editorials that examine matters on which local citizens disagree are more vigorous than those on which there is consensus.

3. Argumentation. The vigorous editorial writer will pick a side, lay out its case, and let you know where the newspaper stands.

4. Mobilizing information. People who are persuaded by the argument will need to know what they can do about it.

Two teams of graduate students rated a group of newspapers, one in April 2002 and the other in January 2003. They met the usual tests for intercoder reliability, meaning that the definitions were clear enough so that the two judges agreed on the presence or absence of each element almost all of the time.[16]

Moreover, editorial vigor was stable across the two time periods. Across twenty-nine newspapers that were measured at both points in time, the vigor ratings were positively and significantly (although not highly) correlated.[17]

This analysis used the 2002 measures of vigor which covered a larger group of newspapers and the same dates that were used for the localism measure. This updating turns out to be important, because there is a positive and significant correlation between localism—the proportion of stories produced by the newspaper's own staff—and editorial vigor.[18] In other words, a newspaper that can assemble a staff with the resources to cover local news can also attract vigorous editorial writers.

A red flag should go up here. Can editorial vigor, too, be just a function of newspaper size?

The paper with the most vigorous editorials was also the largest, *The Philadelphia Inquirer.* The least vigorous was a smaller paper, the *Columbus Ledger-Enquirer.* Weighting the five elements of vigor equally and assigning one point to each, Philadelphia editorials averaged 2 on the 4-point scale. The Columbus mean was 0.43. (The mean for all papers measured in 2002 was 1.37.) And the larger markets tended to have the more vigorous editorials, although the effect was not quite linear. As

---

16. All measures attained a minimum value of .7 on the Scott's pi test.

17. The correlation was .464, p = .026.

18. Correlation = .396, p = .021, N = 34.

with localism, market size did not matter once it exceeded four hundred thousand households in the home county. And it mattered most among the smaller markets.

Here's a look at the scatterplot of editorial vigor by market size expressed as the log of number of households in the home county. Some of the newspapers have been identified by city so that you can see the trend. Once again, a log scale is used to keep the papers from clumping together at the low end of the market-size scale.

Figure 7-2: Editorial vigor increases with market size

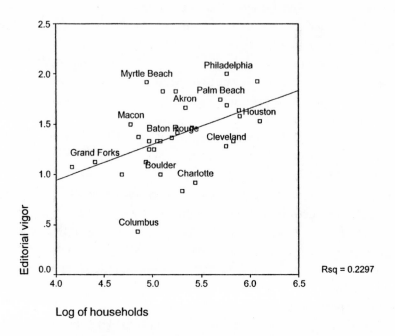

Another advantage of the scatterplot is that you can see who is doing better or worse than their market size would lead you to expect. Those above the line, such as the *Sun-News* of Myrtle Beach, South Carolina, and *The Philadelphia Inquirer,* are vigorous for their size. *The Columbus Dispatch* and *The Charlotte Observer,* unless we caught them in bad weeks, are relatively vapid. In the competitive Akron-Cleveland market, the *Akron Beacon Journal* was more vigorous than its larger neighbor in Cleveland, *The Plain Dealer.*

This variance from expectation can make an interesting variable in itself. It's called the residual because it represents left-over variance in editorial vigor that is not explained by the effect of market size. We can think of it as the vigor score adjusted—or "corrected," if you prefer—for market size.

We can do this correction, but it doesn't help to predict circulation success.

Editorial vigor in its raw form and editorial vigor adjusted for market size yield somewhat different rankings. Philadelphia falls from first to third place, edged out by Myrtle Beach and Greensboro. Columbus, Georgia, remains on the bottom. But in terms of predicting household penetration or penetration robustness, editorial vigor is no help, whether it is adjusted or not.

Illustration and Interpretation

Using a subset of the sample, the same twenty-two newspapers that were in the accuracy survey (chapter 5), my assistants measured the proportion of stories with illustrations and counted the proportion that were more interpretive than straight reporting. The editors had considered both of these factors important. But neither correlated with anything interesting, including circulation success.

If editors were actively managing ease of use, there ought to be a correlation between number of illustrations and the readability scores. There wasn't. However, in this small sample, there was a relationship between illustration and interpretation. Newspapers with lots of interpretive stories had more illustrations.

Quantity of News

Other things being equal, a good newspaper should contain more news than a bad one. This is the most intuitive of the quality indicators. For openers, let's look at the relative size of the news hole, expressed as the percent of the paper's printable area occupied by news.

Percent news correlates positively with household penetration.[19] For

19. Correlation = .369, p = .038, N = 32.

those of us who want to make the case that news is good business, it would be tempting to stop right there. But we can't.

On further inspection, it becomes apparent that market size correlates negatively with percent news, and the effect is strong.[20] Bigger markets have lower ratios of news space to ad space. As they take advantage of economies of scale, they use it to increase the amount of advertising more than the amount of news.

As already noted, the larger markets tend to have lower household penetration. Partial out the market-size effect, and the positive relationship between percent news and penetration disappears into the mist. There is some room for argument here. Maybe the smaller relative news hole is the reason large markets have lower household penetration, and the problem could be fixed with a larger investment in news. But the intuitive path of causation is that market size independently affects both penetration (big markets are harder to penetrate) and relative size of the news hole (big markets have a smaller percent news). The lower penetration in larger markets is due to their greater distribution problems and a less cohesive public sphere caused by greater population diversity. Larger markets also have more distractions in the form of alternate news outlets. I am inclined to give priority to market size in this causal chain because it is a relatively sticky variable, not as subject to change as news hole or penetration. So my tentative call is that the seeming effect of news hole on penetration is only an illusion.[21]

Want to prove me wrong? Here's something to try. Segment a metropolitan market into tight zones for both news and advertising to create a series of cohesive public spheres. Create influence in each of these spheres and reinforce it with the metropolitan influence of the mother paper. Do that in enough markets, and we might see some convincing evidence that the disadvantages of size can be made to go away.

On to absolute size of the news hole. As with other quality indicators, this one is strongly related to market size and forms a nice logarithmic curve with a break at about four hundred thousand households.[22]

20. Correlation = -.568, p = .001, N = 32.

21. Controlling for log of households, the correlation between percent news and penetration is -.0471, p = .802.

22. Correlation with log of households = .907, p = .0000000000008 (about one in a trillion), N = 32.

Figure 7-3: Absolute news hole by households in home county

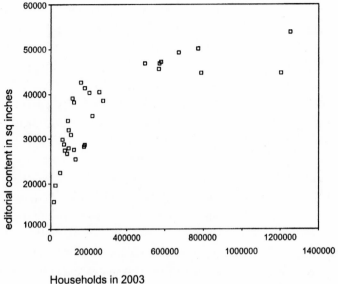

Households in 2003

As with relative news hole, we are in danger of being fooled by the data unless we partial out the effect of market size on the absolute news hole. When we do, the result is discouraging. Size of the news hole, adjusted for market size, has no visible effect on penetration, robustness, circulation, or readership. The correlation is positive in most cases, but it never approaches statistical significance.

Before giving up on content, let's try one more thing. Following the good example of Lacy and Fico, I combined several individual measures of quality, including localism and editorial vigor, into a simple additive index. Maybe their collective power will tell us something that individual measures can't convey.

Lacy and Fico corrected for market size by including city population in a regression equation with their index. I used a different strategy, correcting each item individually and using number of households in the home county as the indicator of market size.[23] It shouldn't matter very

23. The correction was made by regressing each quality indicator against market size and saving the residuals.

much if the underlying theory—that quality causes circulation success—is strong.

Like Lacy and Fico, I used standardized scores for my index.[24] So what is the effect of this composite measure on circulation, penetration, and robustness? None. No matter how you slice it, the effect just isn't there. Most of the correlations were negative, and all were so close to zero that it didn't matter. None came close to statistical significance.

Why this failure to confirm Lacy and Fico's perfectly reasonable and intuitively satisfying finding that quality is associated with better circulation performance?

A major difference between their larger and more carefully selected sample and mine is in the initial correlation between quality and market size. For Lacy and Fico, market size explained less than half the variance in quality, which left some room for other factors to show an effect. In my case, using the log of households in the home county as the size indicator, and home county circulation penetration as the test of business success, size explained most of the variance in the circulation indicators. When I adjusted for market size, there wasn't much variation left for the quality indicators to explain. Putting it another way, the newspapers all looked pretty much alike once the variation for market size was taken out, which makes it more difficult for a correlation procedure to predict anything.

What happened? By focusing solely on home counties, I might have missed the places where quality has the most measurable effect. Perhaps it matters most in the more marginal areas. By focusing on weekdays, I might have missed an important quality effect in the Sunday papers. This is not welcome news. It would be so much better to find that quality has a robust, across-the-board effect that shows up wherever and however you look. We are less interested in effects that need a lot of excuses for not showing up.

Let's consider another possibility. Maybe the world has changed.

Lacy and Fico were working with content measures from November 1984 and circulation figures from about that same time.[25] Their sample

24. Each value is expressed in standard deviation units above or below the mean for all cases. For example, for a given newspaper, a standardized score of 1 indicates that its vigor is one standard deviation above the mean for all newspapers in the study.

25. Lacy and Fico cited the 1985 and 1986 *Editor and Publisher* Yearbooks as their source of circulation data. Because of lags in the ABC reporting process, the published

was carefully drawn to represent a variety of competitive situations: monopoly, including dual ownership; joint agency; and competitive newspapers. My convenience sample, with the Knight Foundation communities at its core, includes some joint agencies (Detroit) and dual ownerships (Philadelphia) but no newspapers with intracity competition. There are hardly any left.

It's true that there are metropolitan areas where papers in different counties compete at the margins: Miami–Fort Lauderdale, Raleigh–Durham, Akron–Cleveland, to name just a few. And this competition is worth a study in itself, because it is in those border-warfare areas that the effects of quality are most likely to appear in a way that can be measured.

But that kind of competition is sissy stuff compared to what was happening back in the early 1980s. Then there were a number of cases of newspapers struggling for dominance in their market, and using news-editorial quality as a weapon of destruction. In *The Philadelphia Inquirer* newsroom, war metaphors were common as editors plotted the demise of the *Bulletin,* which finally fell in 1982. Where such battles went on, advertisers and readers alike gravitated toward the winning paper, as the loser spiraled down in quality and circulation.

So it's possible that there were more inferior, failing newspapers in November 1984, enabling Lacy and Fico to tap a more diverse and interesting sample than was available in 2002. Add the fact that the Knight Foundation communities are heavily weighted toward ownership by a pretty good, centrally managed group, Knight Ridder, and we have a situation where the quality differences across newspapers are minimized. To show that quality has an effect, we need papers that vary in quality. Those in the current sample do vary, but perhaps not enough. Perhaps quality has to deviate a lot—for better or for worse—to make a measurable difference.

If true, this requirement would be bad news for editors who are hoping to make a difference with small, low-cost changes in quality.

Or, and this is an important possibility to consider, perhaps Bogart was right in 1977 when he tried to tell ASNE members that they were

---

numbers usually represent circulation from the previous year or earlier. More recently, Lacy has been working on an update of that study with circulation data from later years (Steven Lacy, e-mail correspondence, December 23, 2003).

off the hook. Maybe the readership decline was not then, and is not now, the fault of the editors.

People in newsrooms—and I have been one—tend to overestimate the effects of their work. The notion of media effects as swift and powerful was nurtured by such exceptional events as William Randolph Hearst promoting the Spanish-American War and the famous 1938 Orson Welles broadcast of *War of the Worlds*. But the first systematic research in the 1940s, by Paul Lazarsfeld and others on the effect of media on voting behavior, led to the opposite theory, that media effects are minimal.[26]

The current consensus is somewhere between. Media effects exist, but they are subtle. They cultivate cultural change, they set political agendas, but they act slowly and quietly. Like the slow-moving migrants from the Bering land bridge, we're not aware of the creeping consequences.

And so it could be with the effect of quality news and editorial content on the business of newspapers. Tracking it over a long period of time, we could tease out the effects and even put a dollar value to them. But one-shot measures, this one included, are not adequate. For vindication of the belief that editorial talent and social responsibility yield profitability, we must look beyond cross-sectional studies of daily content and turn to process measures.

In the next chapter, we'll look at a key player in the process, the copy editor, the last line of defense in newspaper quality control.

26. Paul F. Lazarsfeld, Bernard Berelson, and Hazel Gaudet, *The People's Choice: How the Voter Makes Up His Mind in a Presidential Campaign*, 3rd ed. (New York: Columbia University Press, 1968).

8

The Last Line of Defense

W H E N Christine Urban reported on newspaper credibility in 1989, she gave equal weight to factual errors and mistakes in spelling or grammar as sources of public mistrust.[1] In chapter 5, we found support for her assertion that factual errors are important. Now we turn to spelling and grammar.

Copy editors are the last line of defense in protecting the newspaper from error. They have more control over spelling and grammar than they do over factual error. Beyond verifying names and addresses, newspapers do not routinely fact-check their writers.[2] A copy editor who knows the community well might question a reporter's less intuitive assertions, verify a fact in the newspaper's archives, or even call a source to check the spelling of a name. But the main concern is with form.

In a 2003 survey, Frank Fee and I discovered that copy editors representing no more than 15 percent of daily newspaper circulation in the United States agree with the state-

1. Urban, *Examining Our Credibility,* 5.
2. One exception is *USA Today,* which employs a fact-checker to vet the work of freelancers who contribute to the editorial pages.

ment, "My newspaper rewards copy editors who catch errors in the paper." While catching errors before they appear in the paper is of understandably higher priority, it shows that copy editors are considered more as pre-production processors than as quality monitors.

Spelling and grammar are important in literate societies because consistency makes the task of reading easier. Seeing a misspelled word in print can stop you in your tracks and break your concentration. For this reason, newspapers go beyond the guidance found in dictionaries and create their own stricter set of standards. A dictionary is intended to describe how the language is actually used, and so it is tolerant of variant spellings and changes in usage. A newspaper style book prescribes much more specific rules for spelling and grammar. The basic style book in the newspaper business is the one produced by the Associated Press. Since most major newspapers are its clients, it serves as the default standard, and all the others are variations on it.

In 1989, Morgan David Arant and I realized that an electronic database could be employed for a task its designers never intended: finding errors in spelling and grammar and comparing their rates in different newspapers. Electronic archives were still a novelty, but we were able to search fifty-eight newspapers archived in the DataTimes and VU/TEXT systems that were dominant at the time.

Our technique was simple. We searched for the following errors:

> miniscule (instead of minuscule)
> judgement (instead of judgment)
> accomodate (instead of accommodate)
> most unique (instead of unique)

The misspellings "miniscule" and "judgement" have become so common that dictionaries are starting to recognize them. Newspaper style books do not. The absence of the second "m" in accommodate is not allowed by dictionaries. And "unique" by definition is not modifiable by a superlative. A thing is either the only one of its kind or it is not.

To allow for differences in newspaper size and varying word frequencies, we expressed each misspelling or misuse as a proportion of the total—correct and incorrect—uses. We found that "minuscule" was misspelled 20 percent of the time across all 58 newspapers. The unwanted "e" in "judgment" and the missing "m" in "accommodate" accounted for 2 percent of all usages of each word. And the inappropriate "most" was attached to 1 percent of all uses of "unique."

Because databases were new, our sample then was limited to a six-month period in the first half of 1989. The four items were nicely inter-correlated, a sign that they measured some underlying factor that we chose to identify as editing accuracy.

To adjust for the differing frequency of the test words, we standard-ized the scores before ranking the newspapers. In other words, the error rates were expressed in terms of their standard deviations from the av-erage across all fifty-eight newspapers.

The Ghost in the Newsroom

The best-edited of the fifty-eight papers was the *Akron Beacon Jour-nal,* scoring nearly a full standard deviation above the mean. I called Dale Allen, the editor, to congratulate him and ask how that feat was managed. He attributed it to corporate culture.

"That's the residue of Jack Knight spending six months of the year here," he said. "He hated stuff like that." Knight had been dead for eight years. The culture that he created was persisting.[3]

The lowest-ranking paper on the list was *The Capital* of Annapolis, Maryland, seriously lagging at 2.6 standard deviations below the mean. Its explanation was equally simple: the newspaper had not yet invested in a spell-checker for its electronic editing system.

Most papers did use computer spell checking back then. Errors still got into the paper. How? The spell check is not the last operation before a story goes into the paper. Last-minute changes can involve the intro-duction of words and sentences between the spell check and making the plate for the printing press. Sometimes the copy editor will intro-duce the error.

When Arant and I reported our results, we warned future researchers that electronic librarians might forestall efforts like ours by fixing spell-ing errors in the archives. In most cases, that hasn't happened, although

3. Philip Meyer and Morgan David Arant, "Use of an Electronic Database to Evaluate Newspaper Editorial Quality," *Journalism Quarterly* 69:2 (Summer 1992), 447–54.

the Lexis system incorporates an automatic spelling fixer that hides obvious errors such as "accomodate."

To see how editing accuracy compares with reporting accuracy in making newspapers credible, I set out to replicate our methodology for the twenty-two newspapers for which Scott Maier and I had collected data on reporting accuracy. Easily accessible databases were available for twenty of them (*The Detroit News* and *The Herald-Sun* of Durham, North Carolina, were the exceptions), and I engaged a world-class newspaper librarian, Marion Paynter of *The Charlotte Observer,* to do the searching. She used Dialog for seventeen of them and Factiva for three.[4]

Instead of limiting the search to six months as Arant and I had done, we used the period 1995 to 2003, roughly the span for which I had been tracking circulation success.[5] And a broader selection of errors was employed. Here is the list, along with the overall frequency of each:

Table 8-1

Error	Rate	Base
miniscule for minuscule	12.2%	4,163
supercede for supersede	11.3	1,167
underway for under way	10.8	51,647
alright for all right	03.4	31,412
general consensus for consensus	02.3	33,039
innoculate for inoculate	01.8	739
judgement for judgment	01.7	48,671
most unique for unique	01.3	79,347

Now it is true that these usages might appear legitimately, as, for example, in a story about misspelling, or in an exact quote from a news source who calls something "the most unique." And the AP stylebook

4. There was no visible systematic difference between the two systems. The three newspapers with Factiva archives were near the top, middle, and bottom in the rankings of editing accuracy.

5. The starting month was January 1995, with the following exceptions: *Boulder Daily Camera,* March 1995; *Duluth News-Tribune,* November 1995. All finished in November 2003.

authorizes the use of "underway" as one word in those rare cases where it is used as an adjective before a noun in a nautical sense: "the underway flotilla."[6] But these instances are so rare that I judged their influence, if any, to be random.

The search of the twenty newspapers yielded a total of 250,158 word finds of which 3.95 percent were errors. It's not clear how these results compare to the 1989 study. "Minuscule" was misspelled much less often this time—12 instead of 20 percent of all attempts—but there is no way to tell whether spelling has improved or whether the current sample of twenty newspapers is of better intrinsic quality than the previous sample of fifty-eight. "Most unique" appeared with about the same frequency in 1989, while "judgment" with the unwanted "e" added was somewhat more common back then—2.2 percent of all attempts compared to the current 1.7 percent.

The different samples at different times are close enough to give us some reason to believe that these are fairly stable measures, not something idiosyncratic and meaningless. It really is a way to measure editing quality.

There are two ways to analyze these data. First, let's look at it from the reader perspective and just count the raw percentage of errors. The more often a spelling or usage is wrong or inconsistent, the greater the burden on the reader. So for our first look, here is the ranking of the twenty newspapers according to the number of errors as a percent of all uses. The best are listed first.

Here's what this means to a reader. If you are perusing a random issue of the Boulder *Daily Camera* from the study period and you encounter one of the eight words, chances are better than one in ten that it will be wrong. At the *San Jose Mercury-News*, the risk is closer to one in a hundred. That's variance.

But to really be interesting, variance has to correlate with something. If editing accuracy is an indicator of general newspaper quality, then it should predict all sorts of things, including reporting accuracy, credibility, circulation penetration, and robustness.

It doesn't. Looking at all the relevant scatterplots and correlations, no

6. Norm Goldstein, ed., *AP Stylebook and Briefing on Media Law* (New York: Associated Press, 2000), 255.

Table 8-2

Newspaper	Error rate
San Jose Mercury-News	1.14%
Lexington Herald-Leader	1.95
Raleigh News & Observer	2.35
Akron Beacon Journal	2.37
Detroit Free Press	2.50
Cleveland Plain Dealer	2.58
Wichita Eagle	2.68
St. Paul Pioneer-Press	2.76
Duluth News Tribune	3.10
Grand Forks Herald	3.69
Charlotte Observer	4.06
Palm Beach Post	4.25
Philadelphia Daily News	4.29
Columbus Ledger	4.33
Tallahassee Democrat	4.61
Philadelphia Inquirer	4.65
South Florida Sun-Sentinel	4.91
Aberdeen American News	6.66
Miami Herald	7.77
Boulder Daily Camera	11.08

interesting pattern turns up. Whatever readers want in a newspaper, spelling accuracy appears not to be a primary concern. I can think of two possible reasons, one of which requires further exploration.

The first possibility is that the overall standard for newspaper editing in the United States is so high that readers feel no cause to complain about the exceptional minor lapse. Whether you write "miniscule" or "minuscule," the reader is going to know what you mean. It might bring a few picky old journalism professors like me up short, but most readers won't notice, and those who will don't care.

Standardization in spelling is a relatively recent development. Colonial printers were content to get the phonetics right, and with information still a scarce good, readers were so happy to be getting it that the fine points of presentation didn't matter much. This laxness applied even to proper names. In searching my own family history, I found that I come from a long line of bad spellers, because there were eighteenth- and nineteenth-century ancestors who freely interchanged the spellings of "Meyer," "Myer," "Meier," and "Maier." Sometimes different spellings referred to the same person in the same document.

So you could make an argument that spelling is not important. But we don't. One notably bad speller at the *Miami Herald* in the 1950s was the late Tom Lownes. His spelling was so bad that Al Neuharth, then the assistant managing editor, sat him down for a talk.

"Every newsroom needs one bad speller," Neuharth said. "We need that person to hold up as an example for the rest. He can be a lesson for all the others. We can say to the staff, 'Look at this person, a terrible speller, a sad case. You don't want to be like him.' Having such a person is useful, and we appreciate it."

As soon as Tom relaxed visibly, Al added:

"In this newsroom that person is Steve Trumbull. You learn to spell."[7]

There is another possible explanation for the negative findings about the effect of bad spelling. Look at Table 8–1 again. Two of the most commonly misspelled words, supersede and minuscule, have relatively rare occurrences. "Unique" is used far more than the others, but misuses are rare. By basing the evaluation of newspaper editing on raw frequency, we are giving disproportionate weight to "unique" and hardly any weight to the two most-abused words.

That's fine if effect on the reader is what we are looking for. But if we want to measure copy-editing skill, the eight test words should be given equal weight. A simple statistical trick, using standardized scores, makes that possible.

But first, let's pull one more trick out of the bag. Factor analysis is a tool invented for psychology to find the underlying elements influencing responses to a large number of test items. Applying it to these eight words reveals that they form two logically distinct groups.

The first group includes the words that are pure misspelling rather than matters of style and grammar: all right, minuscule, judgment, inoculate, and supersede. Together, they form the cleanest test of spelling skill.

7. Tom Lownes, personal conversation, Miami, ca. 1959. Quoted from memory.

Two of the remaining three are more indicative of grammar than spelling issues: unique and consensus. For reasons that are not entirely clear, newspapers that get those wrong also tend to mess up "under way." So let's treat those three in a separate index that measure grammar and style more than pure spelling.

By calculating the standardized scores (where the mean is 0 and each newspaper's score is the number of standard deviations away from the mean—positive if above, negative if below), we give each item equal weight regardless of the number of times it turns up in the newspaper.

Copy-Editing Skill

That leads to results less meaningful to readers, but it is a better test of editing skill. Perhaps it will be correlated with something interesting.

The rankings of the newspapers on these new dimensions are in Tables 8–3 and 8–4. The best-spelling editors are at the *San Jose Mercury News,* which is more than one standard deviation below the mean in the standardized error score.

Just by eyeballing the list, we can see that it correlates with something interesting: market size. Papers in larger markets are found near the top of the list and those in small places are closer to the bottom. Larger papers have more resources because of their economies of scale. They can afford to have more (and better) copy editors than smaller papers.

An empirical check is possible because of the availability of newsroom census data from the American Society of Newspaper Editors. Sure enough, there is a statistically significant correlation between number of copy editors and spelling accuracy. The number of copy editors explains 25 percent of the variation in spelling accuracy across the newspapers.[8]

There are some interesting outliers. One pretty big newspaper with lots of copy editors ranks low on spelling accuracy. Another paper with less than half as many copy editors has one of the better spelling scores.

If partial correlation is used to take out the effect of market size, number of copy editors still has a strong effect that leans toward statistical significance.[9] But when the direct effect of circulation size on spelling errors is sought—with number of copy editors controlled—

8. $R = -.497$, $p = .026$.
9. Partial $R = -.423$, $p = .071$, controlling for daily circulation in 1995.

Table 8-3: Newspapers ranked by spelling skill

Rank	Newspaper	Error score
1	San Jose Mercury-News	-1.107
2	Philadelphia Inquirer	-.999
3	Wichita Eagle	-.927
4	Miami Herald	-.869
5	St. Paul Pioneer Press	-.435
6	Lexington (Ky.) Herald	-.407
7	Cleveland Plain Dealer	-.384
8	Detroit Free Press	-.335
9	Akron Beacon Journal	-.235
10	Grand Forks Herald	-.059
11	Philadelphia Daily News	.081
12	Charlotte Observer	.157
13	Duluth News Tribune	.176
14	Boulder Daily Camera	.203
15	Palm Beach Post	.207
16	Raleigh News & Observer	.345
17	South Florida Sun-Sentinel	.556
18	Tallahassee Democrat	.645
19	Columbus Ledger	1.193
20	Aberdeen American News	2.269

Based on minuscule, supersede, inoculate, all right, and judgment

there is no effect at all. So newspaper size is not really an issue except to the extent that it enables the publisher to hire more copy editors.[10]

The ratio of copy editors to reporters, found to be important when we looked at math errors in chapter 5, has no measurable effect on spelling accuracy. Yet, absolute size of the copy desk does have an effect, perhaps because it increases the interaction and reinforcing behavior of the copy editors. Perhaps the size increases the esprit de corps and morale. Or maybe it just means that copy editors have more time to spend on each story. (I'll have more to say about that.)

10. Partial R = .149, p = .542. Note the sign change. With number of copy editors controlled, error increases (but not significantly) with circulation size.

Spelling skill appears to be fairly stable over time. A before-after comparison is possible for twelve of the twenty newspapers because they were part of the Meyer-Arant list of fifty-eight in 1989. The standardized scores were positively correlated across the years with 1989 results explaining 31 percent of the variation in the 1995–2003 period.[11]

Rankings among the twelve changed drastically for three papers. The *Akron Beacon Journal* fell eight places, from first to ninth, Jack Knight's ghost evidently having vacated the premises. The *Charlotte Observer* fell seven places, from fourth to eleventh. And *The Miami Herald* improved by six places, climbing from ninth to third. All of the other newspapers on the list remained within one or two places of their 1989 rankings.

There is more to editing than spelling, of course. We turn now to the three often-misused words that were measured in addition to the five spelling examples. They are not highly intercorrelated, so there is not much of a case to make that they are indicators of the same kind of editing skill. That needn't stop us from using them in an index because a newspaper is still better off getting them right. Using the standardized scores for success at avoiding "underway," "most unique," and "general consensus," we get the set of newspaper rankings in Table 8-4.

There is some correlation with the spelling list, but not a lot.[12] The *San Jose Mercury News* tops both lists. Could this be a consequence of having Knight Ridder corporate headquarters in San Jose? *The Miami Herald* is inconsistent, ranking high on spelling but low on grammar. The *Daily Camera* is at the bottom by virtue of its compulsion to use "general consensus." The word "consensus" appeared in the paper 627 times in the study period, and it was linked to "general" on 47 of those occasions.

Let's look now at the characteristics of the copy desk that might enhance or retard its ability to do good editing. What is a copy desk? When Polly Paddock joined the desk at *The Charlotte Observer* in 2003, she provided a colorful definition: "a crack team of eagle-eyed editors

11. R = .554, p = .062, which leans toward statistical significance. Rank-order correlation was also positive with Spearman's R at .503, p = .095. Referring to the range of .05 < p < .15 as a test value that "leans in a positive direction" has been suggested by John Tukey and is discussed in Robert P. Abelson, *Statistics as Principled Argument* (Hillsdale, N.J.: Lawrence Erlbaum Associates, 1995), 74–75.

12. Spearman's rank order correlation = .286, p = .209.

Table 8-4:   Newspapers ranked by grammatical accuracy

Rank	Newspaper	Error score
1	San Jose Mercury-News	-1.03
2	Raleigh News & Observer	-.94
3	Lexington (Ky.) Herald	-.64
4	St. Paul Pioneer Press	-.54
5	Akron Beacon Journal	-.42
6	Cleveland Plain Dealer	-.41
7	Philadelphia Inquirer	-.29
8	Columbus Ledger	-.24
9	Charlotte Observer	-.18
10	Tallahassee Democrat	-.15
11	Wichita Eagle	-.14
12	South Florida Sun-Sentinel	-.08
13	Detroit Free Press	-.07
14	Duluth News Tribune	-.02
15	Grand Forks Herald	.04
16	Palm Beach Post	.33
17	Aberdeen American News	.65
18	Durham Herald-Sun	.78
19	Miami Herald	.80
20	Philadelphia Daily News	.82
21	Boulder Daily Camera	1.74

who ferret out misspellings, errors of grammar and syntax, muddled thinking, potential libel and all other manner of messes that might otherwise find their way into the newspaper."[13]

"Time Is Quality"

Copy editors don't get the glory and the bylines enjoyed by reporters, but their skills are in great demand, and their starting salaries

13. Polly Paddock, "Writing My Life's Next Chapter," *Charlotte Observer*, July 20, 2003.

are generally higher than what is needed to attract reporters. They don't meet as many interesting people, but they don't have to go out in bad weather either. Their job can be frustrating because in addition to the list of responsibilities given by Paddock, the job of page composition, once part of the back shop, has been moved by computer technology to the copy desk. That means many are pressed for time. "To a publisher, time is money; to an editor, time is quality," said John T. Russial in one of the early studies of the effects of newsroom pagination.[14]

To find out what copy editors think of their jobs, Frank Fee, former copy desk chief of the *Rochester Times-Union* and now my colleague at Chapel Hill, designed a survey. I helped with the sample.

Surveys of newspaper personnel are tricky. If you make them representative of all newspapers, you get a lot of small papers because, after all, most newspapers are small. To get around this problem, many academic newspaper surveys are stratified, meaning they draw separately from groups of papers of different size.

Fee and I used a different strategy. We used what sampling guru Leslie Kish has called a "PPS" sample for probability proportionate to size. Each newspaper in the United States that is audited by the Audit Bureau of Circulations had a chance of being included that was proportionate to its total circulation size in 2002.[15]

This procedure means that the largest papers fell into the sample automatically. And it has the advantage of representing newspaper customers rather than the newspapers themselves. In other words, the editors in the sample represent the total audience of people who buy newspapers.[16]

The last line of defense for quality in newspaper journalism is not a

---

14. John Russial, "Pagination and the Newsroom: A Question of Time," *Newspaper Research Journal* 15:1 (Winter 1994), 91–101. See also Doug Underwood, C. Anthony Giffard, and Keith Stamm, "Computers and Editing: The Displacement Effect of Pagination Systems in the Newsroom," *Newspaper Research Journal* 15:2 (Spring 1994), 116–27.

15. Leslie Kish, *Survey Sampling* (New York: Wiley Interscience, 1965), 217–53.

16. The basic sample had a list of 221 newspapers. We merged it with the membership list of the American Copy Editors Society (ACES) to get names of individuals. For newspapers with no ACES members, we wrote and phoned editors listed in the 2003 *Editor and Publisher Yearbook* and asked them to nominate three copy editors for the survey. With this procedure, we obtained names and addresses of copy editors for 80 percent of the newspapers in the sample.

happy place. John Russial has noted that when computerized page composition moved work from the back shop to the copy desk, the newsroom did not get proportional extra staffing. Other responses to competition, including zoned editions and feeding copy to online editions, put still more pressure on the copy desk.

Fee knew from personal experience the kinds of things that bug copy editors, and he included a long list of them in the questionnaire as an agree-disagree list. There was much more disagreement than agreement on the following (numbers show percent agreement and disagreement; they don't add to 100 because some responses were neutral).

- My newspaper rewards copy editors for good story editing (Disagree by 68 to 10).
- My newspaper rewards copy editors who catch errors in the paper (Disagree by 63 to 14).
- Proportionate to their numbers in the newsroom, as many copy editors went to professional conventions and conferences last year as reporters (Disagree by 61 to 18).
- Copy editors at my paper have as much power as reporters to shape content and quality (Disagree by 47 to 30).
- Most copy editors feel appreciated at my newspaper (disagree by 47 to 21).
- Copy editors at my paper are held in the same esteem as reporters (Disagree by 40 to 17).

These six items make the kind of scale that statisticians love because of their strong intercorrelation—a sign that six items, despite their diverse content, are tapping the same underlying factor. We get to name it whatever we want, so let's call it "respect." Copy editors who feel respected will have higher agreement scores on these items.[17] The mean respect score is 2.37—on a scale where 1 is the minimum, 3 is neutral and 5 is the maximum.

The mail survey was conducted in the fall of 2003 and yielded a response rate of 73 percent of all those invited to participate. Of the original list of newspapers, 76 percent were represented in the final survey. To maintain the PPS feature, weights were added to give each of these newspapers the same representation whether the number of editors responding was one, two, or three. The final Ns were 337 editors and 169 newspapers.

17. Chronbach's Alpha for the six-item scale is .741.

Now comes the question we've all been waiting for. If copy editing makes a difference in the newspaper's business success, it stands to reason that those news organizations that treat their editors with respect will do better in resisting the long-term decline in circulation penetration.

They do, and the difference leans toward statistical significance.[18] Newspapers whose copy editors score a two or better on the five-point respect scale hung on to an additional 1.5 percentage points of home county penetration between the 2000 and 2003 ABC reports. (This is the difference after a few papers with extreme shifts due to unusual local situations or changes in the method of counting have been eliminated.)[19]

To make this perfectly clear, here's the same information expressed in a different way: In the period 2000–2003, the low-respect papers kept 96.8 percent of their home county circulation penetration from one year to the next. The high-respect papers kept 97.3 percent. That half-percentage point difference per year adds up mighty fast.

This leads us to wonder what causes copy editors to feel respected. The biggest factor uncovered in the survey is a light work load. Fifty-nine percent of the editors reported having processed fourteen or fewer stories in their last shift. Their mean respect score was 2.43, well below the neutral point of 3.0. But it was worse for those who had to work more stories.

The harder-pressed copy editors had a collective respect score of 2.26, a difference that is significant in the traditional sense.[20] And they were unhappier on a broad array of factors. They felt less respected by their newspapers' reporters, they saw fewer opportunities for professional development, they liked their bosses less, and they were less likely to feel rewarded for good work.

Here we have strong support for Russial's finding that the new technology-imposed demands on the copy desk are creating some pressure in newsrooms. The Fee survey indicates it is also creating some foul air. Even though this development is relatively new, its effect is already working its way down to the bottom line enough to show a clear hint of a difference. Attention should be paid.

18. Student's t = 1.810, equal variances assumed, p = .071.
19. Outliers and extreme values were identified using the SPSS Explore procedure. Seventeen newspapers were eliminated for this part of the analysis.
20. Student's t = 1.977, equal variances assumed, p = .049.

9

Capacity Measures

M E A S U R I N G quality in journalism is a little bit like measuring love. Once at a meeting of professors who specialize in media ethics, I expressed the view (borrowed from science fiction writer Robert A. Heinlein) that if a phenomenon exists, it has to exist in some quantity. And if there is a quantity, there ought to be a way to measure it.

My colleagues, most of whom had formal training in philosophy, were aghast. "You can't measure love," one of them said. "And we know that love exists."

Well, call me conceited, but I think I can measure love. It's interesting, it affects human behavior, and that behavior can be visible enough to measure. True, it wouldn't be a direct measure. But in social science, we often are quite content with indirect measurement. Arriving at such a measure is called operationalization. We could talk forever about the true definition of love, but our operational definition would deal with some aspect of love that surfaces enough in the visible world to create a measurable effect. The Gospel According to St. John, for example, treats love as a continuous variable and offers an operational measure for the end point on the continuum: "Greater love hath no man than this, that a man lay down his life for his friends."[1]

Newspaper quality also has its visible manifestations. So far, we have dealt mostly with content measures, for example, the ratio of news to advertising or the accuracy of the staff-written stories. But we can also measure the consequences of that content by going directly to the audience through survey research and finding out how much they read, trust, and respect the newspaper.

This chapter is about yet a third way: measuring the capacity of the newspaper. Think of the newspapering process as a black box. Capacity is what you put into the box. Content is what is created inside the box. Consequence is what comes out of the box. I like to think of it as the 3-C model: capacity, content, and consequence.

The leadership role in measuring capacity was assumed by the Poynter Institute when it hired Rick Edmonds, a freelance investigator and analyst, to look for operational measures of journalism capacity. He found some excellent sources: the summary reports of the Inland Daily Press Association and the staffing surveys of the American Society of Newspaper Editors.

Capacity is a topic dear to the hearts of newspaper editors because it defines their constraints, the tools they are given by their publishers to create the best output their talent and ingenuity can produce. For this reason, it is also an emotional topic. Both editors and publishers are reluctant to release the intimate details of their businesses, and so the data that Edmonds had to work with were encumbered by the need for secrecy.

Where capacity involves a particular job function, however, it is easier to measure. The presence of an ombudsman, for example, indicates a highly specific kind of capacity, a reader advocate who is responsible for the newspaper's public self-criticism. A similarly specific kind of capacity is found in training. A newspaper with a full-time person responsible for staff training has a different kind of capacity than one that does not.

For a look at the grossest measure of capacity, ratio of staff size to cir-

1. John 15:13.

culation, my assistant Minjeong Kim and I took a cue from Edmonds and used data collected by the American Society of Newspaper Editors. Most of the data used in this book are publicly archived and available. The ASNE data is not public because it is collected with a qualified promise of confidentiality as part of its long-range program for increasing the amount of minority representation on newspaper staffs.

Minority representation on the newspaper, as compared to minority representation in the community, could itself be taken as an indicator of quality. But for now, we'll just look at the raw figures.

To start, we need a benchmark. According to newspaper folklore, a good newspaper employs about one news-editorial staff member for each thousand in circulation.

The ASNE censuses make it possible to track change over time because they have been collected every year since 1978 as part of the organization's specifically quantitative goal to have minority staff reach the same proportion as minorities in the population served. The original target for reaching that goal was the year 2000, but it was later extended to 2025.

While ASNE publishes the minority percentage figures for each participating newspaper on its Web site, it does not release the raw numbers from which those percentages are derived.[2] However, it did provide raw data for the years 1995 and 2000 to a few researchers on condition that values for individual newspapers not be revealed.

For Kim and me, the first step was to merge the ASNE data files with circulation numbers from the Audit Bureau of Circulations. Like the U.S. Census, ASNE has some coverage problems, and not every member newspaper responds every year. Also, many smaller newspapers do not belong to ABC and were excluded from our study for that reason.

We made one other exclusion. The national newspapers, *The New York Times, USA TODAY,* and *The Wall Street Journal,* have economies of scale that make them potentially different from local newspapers. Using some standard statistical definitions, we determined that they are outliers in fact as well as in theory. Dropping them left us with a convenience sample of 477 ABC newspapers that responded to ASNE in 1995 and 616 responding in 2000. We checked the 1995 sample to see

2. http://www.asne.org.

if the conventional-wisdom prediction of one staff member per one thousand circulation was accurate.

It was. The mean news-ed staff ratio for 1995 was 1.04. But there was variation around that mean, and the average grew with the robust economy of the last half of the decade. For the year 2000, the ASNE survey showed the staffing rate had ballooned by nearly a fifth, to 1.18 per thousand circulation.

And there was some variation by circulation size, suggesting modest economies of scale. We divided our sample into four circulation categories and compared news-ed people per thousand circulation in each:

Table 9-1

| | *Staff per thousand circulation* | |
Circulation	*1995*	*2000*
0-15,000	1.15 (N=114)	1.35 (N-182)
15,001 – 150,000	1.05 (N=302)	1.15 (N=369)
150,001 – 300,000	.86 (N=38)	.98 (N=41)
>300,000	.72 (N=23)	.81 (N=24)

While newspaper companies prospered in this time period, the most obvious cause was their ability to raise prices in good times while costs were declining.[3] But since these trends affected everyone in the business, we wondered if there were some small increment of business success that could be attributable to editorial staff size.

Looking at the 473 ABC newspapers that reported to ASNE in both 1995 and 2000, we grouped them into three categories depending on whether they failed to increase staff size in that period, increased staff by up to 10 percent, or increased staff by more than 10 percent. The relative frequency of the three groups:

No gain	35%
Small gain	23
Large gain	41

3. Newspaper Association of America, *Facts About Newspapers 2001* (Vienna, Va., 2001).

Those in the first group, the newspapers that maintained or reduced staff, lost significantly more circulation than the others. In the 2000 ABC county penetration report, they had an unweighted mean circulation that was 93.5 percent of the circulation reported in 1995.[4] Those with small staff growth and large growth alike retained 97 percent of their five-year earlier circulation.[5]

A Reinforcing Loop?

Of course, we have no way of knowing which came first: the staff loss or the circulation decline. We can make a theoretical case for either one or the other as the primary cause. Or it could be a reinforcing loop where lost circulation creates financial pressure to cut staff, which degrades quality and leads to further circulation loss.

We can get a clue—but only a clue—to the primary direction of cause and effect by following what happens over time. To do this, Kim and I chose the same dependent variable, percent of 1995 circulation retained in 2000.

The dependent variable was the news-ed staff per one thousand circulation in 1995 regardless of whether or how that figure changed in the ensuing half decade. All we need to know is whether newspapers that started the five-year period with a more robust staff-to-circulation ratio had better results over the course of those years than those starting with less staff. Because time is one-directional, a positive result would buttress the inference that staff size is more cause than effect of healthy circulation.

A possible spurious effect is immediately suggested. Smaller papers, lacking economies of scale, will have more staff per thousand. They might also be more intimately involved with their communities and less at risk for circulation loss. If so, we should introduce a control to compensate for their lesser economies of scale. But a look at the relationship between size and circulation allays this particular concern. Smaller papers are more susceptible to wide swings in circulation, up or down. But, on average, they suffered no greater loss over 1995–2000 than their larger brethren.

---

4. These reports cover varying audit dates, mostly in the second half of 1999.
5. The between-groups difference is statistically significant ($F = 3.683$, $p = .026$).

Figure 9-1

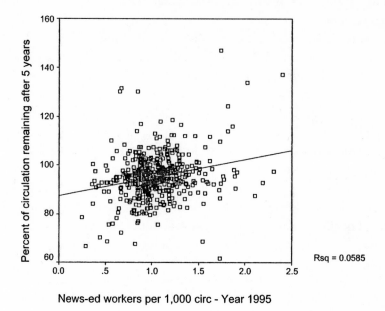

News-ed workers per 1,000 circ - Year 1995

This one isn't as easy to interpret as the ones in chapter 1, so here's a brief explanation. For no very good reason except convention, the horizontal dimension is called X and the vertical is Y. When cause-effect relationships are anticipated, the presumed cause gets the horizontal or X scale. (John Tukey is an exception. All of his plots make X the effect and put it on the vertical, which is consistent with use of X as the unknown in algebra. Tukey was a great innovator and many of his ideas caught on, but not that one.)[6]

Each tiny square on the plot positions one newspaper on the horizontal according to its news-ed staff per thousand circulation in 1995 and on the vertical according to the percent of its circulation retained five years later. If staff size explained everything, the data points would form a straight line sloping upward from left to right. If staff size explains less than that, but still a lot, the points will still look at least a little bit like a straight line. And if it explains nothing at all, they points will form an amorphous blob.

What we have in Figure 9–1 is pretty close to a blob. But a computer

6. For example, Tukey, *Exploratory Data Analysis.*

representation of the best-fitting straight line shows some upward slope. And the computer also tells us that the effect is significant, quite unlikely to be a random quirk, and that staff size explains 5.8 percent of the variation in circulation success.[7]

That's not much. But it's enough to give the advocates of more generous staffing some bragging rights in a business where even small improvements in circulation are cherished.

A much stronger case might be made if we had more data points. The problem with time studies is that causes and effects can persist for long periods. Monday tends to predict Tuesday, last year predicts this year, and how you feel this minute is a pretty good indicator of how you're going to feel a minute from now. In the case at hand, the papers that had short staffing in 1995 might have been that way because of a losing circulation pattern in a prior period. And that losing pattern might have persisted to 2000. Because we only looked at 1995 and 2000, we would have the illusion that the short staffing came first.

Statisticians call this "autocorrelation." The way to deal with it is to collect a number of data points over a considerable period of time. Then you can estimate the effect of autocorrelation with enough precision to make the necessary corrections for it.

The slope of the line shows a rise of 7.427 for each unit increase in staff per thousand circulation. In other words, starting from a base of 87.4 percent circulation retention, a typical newspaper started in 1995 with a staff/circulation ratio of 1 would have been expected to hold on to 94.8 percent of its circulation (87.4 + 7.427). But if its ratio were 1.5 per thousand, the expectancy would be 98.5 percent of its original circulation (everything else being equal).

So how big would the staff have to be to have kept circulation from falling at all? Push the pencil, and the math works out to 1.7 staff per thousand circulation. That's high, but not totally abnormal. The top 9 percent of newspapers reporting to ASNE met that standard in the year 2000. (The confidentiality agreement prevents me from listing them.)

The trouble is that staff size is only one cause in a world where many other causes are competing for management's attention. Just looking at the blob around the straight line on the plot shows us how much

---

7. P < .0005, meaning that an effect of that magnitude has less than five chances in ten thousand of occurring by pure chance.

individual cases differ from what the line predicts. That slope of 7.427 is just an average, and not a good predictor in any single case.

Rick Edmonds tried relating staff size, measured by ASNE's 2000 survey, to a more direct indicator of newspaper quality. He picked a group of five newspapers that he and his colleagues at the Poynter Institute considered "journalistically weak" and compared them to five that were judged to be among the nation's top thirty by editors responding to a *Columbia Journalism Review* poll.[8]

As might be expected from such a small sample, the results were inconclusive. On the whole, the good papers had a 20 percent better staff-circulation ratio, but the entire effect was driven by just four papers, the two best and the two worst.[9] The other six, good and bad, had similar staff ratios. Edmonds went back and enlarged his sample to forty independent newspapers, forty-two in privately held groups and ninety-six in public groups. The results were still ambiguous. When he looked at nineteen of the twenty-one highest-ranking papers in the Columbia poll, thirteen were above average in staffing and six were below. (Two low-staff-ratio newspapers, *USA TODAY* and *The Wall Street Journal,* were excluded because of their size.)

Edmonds also compared similarly small samples of newspapers in publicly owned groups, privately owned groups, and independent (non-group) papers. The conventional wisdom was supported: independent papers had the best staffing, and private chains were better than publicly held chains. But the differences were small. When he looked for evidence of large-scale staff reductions, the results were equally ambiguous. While staff cuts, including layoffs, at publicly owned Knight Ridder had been highly publicized, the publicly owned newspapers among twenty-two that Edmonds looked at actually cut less than those that were privately held. Knight Ridder had been criticized by analysts for having high news-editorial costs, and the cuts simply brought it closer to the industry norm.[10]

Where newspapers are generously staffed, local market structure

8. "America's Best Newspapers," *Columbia Journalism Review* 38:4 (November/December 1999), 14–16.

9. Rick Edmonds, "Lifting the Veil on News Staffing," *Poynteronline* (www.poynter.org/content, retrieved June 5, 2003).

10. Rick Edmonds, "Public Companies No Worse Than Private," *Poynteronline* (www.poynter org/content, retrieved June 5, 2003).

sometimes provides an explanation. A newspaper that addresses separate and scattered audiences with zoned editions needs more staff than one that sends the same message to a compact, contiguous population. A newspaper that is fighting competition will pay for more staff than one that does not feel threatened.

The Fort Worth Case

The most generously staffed newspaper that Edmonds found, Knight Ridder's *Fort Worth Star-Telegram,* was in just such a situation. The Dallas–Fort Worth metropolitan area is one of a handful where two strong dailies face the possibility that economics will eventually leave room for only one of them. (Tampa–St. Petersburg, Raleigh–Durham, and Minneapolis–St. Paul are other examples.) In those situations, the papers can try to observe natural or historical territorial lines. But as their populations grow, they feel the need to contest the newer areas. The winner of those contests might be the one that eventually dominates the market.

In Dallas–Ft. Worth, the contested middle ground is the sprawling suburb of Arlington, which lies south of their shared airport and directly between the two large cities. Edmonds checked it out and found that Dallas had initiated the conflict in the mid 1990s when the *Star-Telegram* was still owned by the Walt Disney Company. The *Star-Telegram* responded with more staff and expanding zoned editions, one in northeast Tarrant County and one in Arlington with the equivalent of seventy full-time staffers in each, including some senior editors and seasoned reporters (a departure from the usual newspaper pattern of sending the greenest journalists to earn their stripes in the suburbs). The case proves once again that newspaper competition can be a good thing for readers.

Edmonds checked it out and found "a big news hole, wholesale local-section zoning, deploying experienced editors and reporters to the two suburban bureaus and making a mini-religion of useful, reader-friendly presentation," as well as increasingly ambitious investigative work. Its 336 full-time professional staff gave it a staff-circulation ratio of 1.5 per thousand. Local coverage is intense with special emphasis on high school sports.

How did the newspaper deal with the damage to profitability? There wasn't any, Edmonds reported in his case study. "Always a good earner, the *Star-Telegram* has increased profits 45 percent over the last five years." In Knight Ridder's internal surveys of reader satisfaction, the *Star-Telegram* finished first three consecutive times, he reported. Publisher Wes Turner won the company's John S. Knight gold medal for performance "that continues to excel by every measure that's important to a newspaper."

The *Star-Telegram*, Edmonds reported, spent 10.5 percent of revenue on news-editorial expenses, a bit above the 9.5 percent reported by the Inland Daily Press Association as average for papers in its size class.

And that opens up another area for possible evaluation of news capacity. The problem is that most newspaper companies don't want to share that data. Edmonds found out just enough to be sure that there is variance. For example, the *St. Petersburg Times,* a highly respected newspaper owned by the Poynter Institute, with its circulation of 337,000 should have been spending 9.5 percent or less because of its obvious economies of scale. Instead, it reported to Edmonds that its figure was 12.5 percent. That's almost a third above the norm.

The Media Management Institute at Northwestern University is working with Inland data, and its work will have great potential if the investigators are permitted to track individual newspapers, their expenditures, and their circulation success over time.

Counting the money and the people are not the only ways to measure capacity. It can also be important to look at the way the people and the dollars are deployed. One important variable is flexibility or what the military calls "surge capacity."

The difference between good newspapers and mediocre ones can be seen any time a major story breaks. Once when a hurricane devastated the North Carolina coast, one major state newspaper sent enough reporters out to provide a mile-by-mile description of the damage, and it paid the cost of extra newsprint to provide enough open pages to tell the story in words and pictures. Another relied mostly on wire services and squeezed all of its reporting into very little more than the normal amount of news space. That gave me the idea to try more systematic comparisons the next time a big story broke. In September 2001, one did.

In fall 2002, some of my students looked at the previous year's 9/11 coverage by various newspapers and found that size was the main predictor of surge capacity. On the whole, larger newspapers tore out the ads at the front of the paper to make room for news. Smaller papers often did not.[11]

One notable exception among the small papers was Knight Ridder's *Centre Daily Times* in State College, Pennsylvania. On the Thursday after September 11, it doubled the size of the paper compared to the previous Thursday and increased the news hole by 269 percent. Some larger papers, including the *San Jose Mercury News* and *The Philadelphia Inquirer,* cleared all the ads out of the front section of the paper to make room for 9/11 coverage.

Specific Indicators

The measures of capacity discussed so far have been quite general. But there is no reason we can't look for highly specific indicators. For example, whether a newspaper has a full-time training director can tell a lot about its commitment to quality journalism. In a profession that is always changing, people on the job need to upgrade their skills, both for their personal satisfaction and to maximize their opportunities.

Here's another example, this one from the distant past. When I was the education writer for *The Miami Herald* in the late 1950s, I was assigned to write a weekly news summary for high school students that could be used in the classroom. It was designed as a quarterfold, meaning that its four pages were the equivalent of one broadsheet page— back when the broadsheets were considerably wider than they are now.

When I first tried to write it, I dutifully assembled all copies of the *Herald* for the previous seven days. Local stories were no problem. Stories that needed following had been followed. But national and international stories were a totally different case. There might be a story about a pending Senate vote on Monday but no follow on Tuesday to tell the reader how it turned out. Wire copy, it turned out, was just being used as filler. The desk didn't have enough personnel to create a national and international report that was coherent from day to day.

I finally had to resort to asking the telegraph editor to save the spiked

11. My thanks to Amanda Crowe, Dale Edwards, Anush Hosvepian, and Stephanie Johnes.

AP copy. That and *The New York Times* made the job possible. But the exercise showed clearly how little thought went into editing the non-local parts of the paper.

Eventually, the *Herald* fixed the problem by adding the position of national editor. That move helped readers and also made life easier for the Washington bureau by giving it a contact person who was paying attention to the consistency of the national report.

In those days, the presence or absence of a national editor would be a good indicator of capacity. It's not today, because national editors are so standard on larger papers that there's no variance to provide comparisons. But there are other positions that are not so common and might be indicators of good journalism.

One of my personal favorites is a newsroom specialist in precision journalism or computer-assisted reporting. The terms used to be interchangeable, but today a computer-assisted reporter can be anybody who substitutes searching the Internet for old-fashioned leg work. By precision journalism, I mean applying social science methodology, including statistical analysis, to the practice of reporting.

It would be fairly easy to evaluate a newsroom's capacity in this area along a cumulative scale. A newsroom with reporters who can do the things deeper on the scale can usually be counted on to have the capacity for the shallower tasks nearer the top.[12]

1. Capture data from press releases, government records, or other Internet sources and use a spreadsheet or database program for sorting and ranking.

2. Make comparisons across subsets of the data with a crosstab program such as SPSS. Merge data from different sources.

3. Obtain and report the results of significance tests and correlation coefficients.

4. Test for causal inferences with three-way crosstabs or multiple regression.

5. Design a sample, then collect and analyze original data.

---

12. In social science, it's called a Guttman Scale. The statistic for evaluating it is the coefficient of reproducibility. It tells you how much knowing the presence of any given item on the scale tells you about the easier items above it.

Here's another position-specific measure worth trying: presence or absence of a reader representative or ombudsman.

The first newspaper ombudsman program was created in Louisville by Norman Isaacs, editor of *The Courier-Journal* when it was a family-owned newspaper. (Gannett bought it in 1986). John Herchenroeder, a former city editor, responded to reader complaints and codified the paper's internal procedures. He also wrote an occasional article explaining or criticizing the newspaper's actions. The idea was copied, but not widely. After reaching thirty-five in the 1980s, the number of newspaper ombudsmen in the United States stabilized at around thirty. Some ombudsmen write columns on a regular basis.

The presence of an ombudsman reflects not only a newspaper's capacity but also an editor's decision on how to deploy that capacity. Among the elite newspapers, *The Washington Post* was quick to create the position, while *The New York Times* declared it unnecessary. The *Times* changed its mind while repairing the damage to its reputation caused by the pattern of intentional deception in its news columns by reporter Jayson Blair in 2003.

This chapter opened with an attempt to demonstrate a link between journalistic capacity and business outcomes, using news-editorial staff per thousand as the capacity measure and the percent of circulation retained over a five-year period as the outcome measure. So I'll close with a narrower example: an attempt to detect the effect of having an ombudsman on a newspaper's ability to hold on to its audience. This one requires a smaller sample, and so I'll use the more specialized measure of circulation, robustness of home county penetration. In the earlier example, hundreds of newspapers were involved, and so individual differences tend to get washed out as alternative explanations for maintaining or lagging circulation. In the smaller sample, it's better to focus on the home county and to count the households so that penetration change over time can be measured. This measure corrects for market size and it uses the close-in circulation as the indicator of the newspaper's overall health.

Minjeong Kim started with the twenty-nine newspapers represented in the membership list of the Organization of Newspaper Ombudsmen (ONO) as of 1996. This was the experimental group. To create a control group, she listed all of the ABC-audited newspapers in order of

circulation. For every paper with an ombudsman, she picked a newspaper that adjoined it in the rankings. To ensure random selection, the
next newspaper higher in circulation was chosen if the ombudsman
paper had an odd-number rank in the total list. When the rank order
was even, the next smaller newspaper was chosen.

The result was two lists of newspapers comparable in size, one with
ombudsmen and one without. Their home county circulation penetration was compared for 1995 to 2000, a period when newspapers across
the country were losing circulation in both absolute and relative terms
(as a percent of households).

The twenty-nine papers with ombudsmen in 1996 had retained, on
the average, 89.3 percent of their home county penetration between
1995 and 2000. The papers of comparable size without ombudsmen
retained only 86.3 percent. The difference of 3 percentage points was
statistically significant.[13]

While our circulation measure takes time into account, the time element is not strong enough to make a firm conclusion about causation.
This caveat is used often in this book, but it bears repeating. These results can be explained in three different ways:

1. The presence of an ombudsman helps readers bond with the newspaper and provides some of the resistance to circulation slippage.

2. A newspaper that is holding on to its readers is more likely to have
the financial resources to afford an ombudsman—considered a frill by
many managements. The ombudsman would therefore be an effect of
business success, not a cause.

3. An ombudsman is just one visible sign of a newspaper that cares
about its reputation—and its influence—in the community. A newspaper that has one is probably doing lots of things right, and they all
add up to an effect on maintaining circulation. (I like this one the best.)

Capacity measures show some promise for large-scale comparative
evaluation of newspapers, but no single one is likely to suffice. Having
a variety of measures and finding measures that are obtainable are both

---

13. Because the list of papers with ombudsmen was a census, not a sample, we used
a one-sample t-test to compare it to the nonombudsmen papers. (p = .011).

important. Until the industry develops more of an attitude of openness—not inappropriate for a communications company—both personnel and newsroom expenditures will probably not be available for large-scale study with the newspaper as the unit of analysis. Aggregate measures might give some clues. They could show, for example, whether circulation success tends to follow or precede cutbacks in staffing in the industry as a whole.

But the best measures will classify individual newspapers in sufficient detail so that effects of investment can be detected. In the meantime, content, because it is manifest and measurable, must remain the most viable indicator of quality.

10

How Newspapers Were Captured by Wall Street

KNIGHT NEWSPAPERS went public on April 22, 1969, with an initial offering of 950,000 shares. John S. Knight had yielded the titles of chairman and chief executive officer to his brother James L. two years earlier, but he retained the title of editorial chairman. In that role, and because of his history as a founder of the company, he was invited to speak at a meeting of Wall Street analysts. He told them, "Ladies and gentlemen, I do not intend to become your prisoner."

In an interview with Dan Neuharth years later, Knight recalled that occasion. "I told them why. I said that as long as I have anything to do with it we are going to run the papers. We are going to spend money sometimes that they wouldn't understand why, for future gains and we did not intend to be regulated or directed by them in any respect." Knight's associates never took him to another analyst meeting.[1]

1. *Akron Beacon Journal*, "John Shively Knight 1894–1981, A Tribute to an American Editor," offprint, June 1981.

It was easier to keep Wall Street at arm's length in those early days because the Knight family retained 58.8 percent of the shares after the initial offering. By the time of Alvah Chapman's tenure as CEO from 1976 to 1988, the company had become Knight Ridder, but still had insider control, even though family was defined more broadly by then.

"We used to have family . . . shareholder meetings," Chapman recalled after his retirement. "It was my family and Lee Hills' family and Bernie Ridder's family and Tony Ridder's family and all the Ridders and all the Knights and the foundation. We used to keep score, and it was like 56–57 percent in my time. It's gone downhill quite a bit from there."[2]

By 2002, institutional and outside investors controlled about 90 percent of Knight Ridder stock, according to P. Anthony Ridder who became CEO in 1995. Contacts with investors and analysts became more intense. "Since I have been CEO," said Ridder, "the number of meetings has increased dramatically, and institutional investors expect you to call on them."[3]

This need to answer to outsiders was a major cultural shock to newspaper companies, which had traditionally been managed in a fairly informal way. This chapter traces that development through interviews with present and former officers of three publicly held companies: Gannett, Knight Ridder, and McClatchy. They have been differently managed, but their three histories have a common thread: their transitions were executed by people who had been socialized to the ideals of social responsibility at newspapers operated by the Knight brothers and their news-editorial guiding light, Lee Hills, in the 1960s and 1970s.

I witnessed a little bit of this history firsthand. In 1962, I left my reporting beats at *The Miami Herald* to become the *Akron Beacon Journal's* correspondent in the Knight Washington Bureau. Edwin A. Lahey, the bureau chief, told me, with a touch of pride, "The Knight Newspapers come together in only two places. One is the Washington Bureau, and the other is Jack Knight's briefcase."

2. Interview with Jane Cote and Philip Meyer, Miami, August 4, 2002.

3. Interview with Jane Cote in San Jose, California, and Philip Meyer, by telephone, October 25, 2002.

Alvah Chapman

It was true. Knight Newspapers consisted of just five newspapers owned by four companies, which in turn were owned by the Knights. "There were minority stockholders in Akron and minority stockholders in Charlotte," Chapman recalled in 2002. Their investments were based on faith in, and personal knowledge of, the Knights.

"They never had seen an operating statement. It just never had been done," said Chapman. There was a board of directors consisting mostly of people who worked for the Knights, and it met just once a year. "They passed out numbers, and they took up the numbers at the end of the board meeting. They didn't let the directors even keep the numbers to ponder over.

"The executive committee was formed the week I joined the company (1960). It was formed because Jim Knight decided to build the Herald building and forgot to tell his brother Jack about it. . . . Jack thought it was going to bankrupt the company. They built a $25 million building. It was just an oversight he forgot to tell his brother about."

Chapman had graduated from the Citadel in 1942 with a major in business and served in the Army Air Corps as a bomber pilot. He learned the newspaper business on family newspapers in Bradenton, Florida, and Columbus, Georgia. "My father was a newspaper publisher. My grandfather was a newspaper publisher. I knew I wanted to be a newspaper publisher when I went to college. I took business because I knew that's what I wanted to specialize in. I knew the journalism thing would come along by absorption."

As the youngest member of Knight's brand-new executive committee, Chapman was assigned to set its agenda. That gave him plenty of freedom to poke around. He found some interesting things, such as $15 million in bank accounts that were not drawing interest—overlooked because of the lack of a consolidated financial statement. Jim Knight believed that *The Miami Herald* never gave refunds when a wrong telephone number or other critical error appeared in a classified ad. Chapman investigated and found that the ad staff kept its records in pencil and just erased an entry to prevent an unjustified charge from being made when an error occurred. He instituted a record-keeping system that let management track errors and know how much they were costing.

Erwin Potts

When Erwin Potts left *The Charlotte Observer* for McClatchy news-papers in 1975, he found a tradition of informality that reminded him of the Knight newspapers a decade earlier. Potts's first job after college had been with the *Charlotte News,* and he never forgot the thrill of his first page-one byline. After service in the Marine Corps, he joined *The Miami Herald* as a reporter at about the same time that Alvah Chapman was walking around with a clipboard, tightening up procedures on the business side. Potts was city editor of the *Herald* when Knight News-papers went public, and Chapman, then executive vice president under Jim Knight, brought him downstairs to be his assistant. The deal was that Potts would go to Tallahassee to be general manager of the *Demo-crat* if his business-side stint worked out. It did. He later moved on to Charlotte, where he was serving as general manager when Knight and Ridder merged.[4]

The nature of the company changed in that time period. Part of it was the greater administrative structure required by the company's size, and part was the increased centralization and command-and-control structure put in by Chapman. The effects of going public were starting to kick in, especially when a recession in 1973–1974 led to demands from headquarters for cost-cutting. Potts was thinking that the newspa-per business wasn't as much fun as it used to be when he got a call from a headhunter working for McClatchy. The company was going outside to look for professional management. Out of curiosity, Potts went to Sacramento to meet the family.

"Eleanor McClatchy was in her seventies then, and she was still in charge of the company. And it was one strange company. It was weird. Totally private. Eleanor was the boss. She ran everything. Nobody crossed Eleanor. She was a sweet wonderful little lady, like your favorite aunt. But for thirty-some years she had been running that company. Three newspapers, seven radio stations, and two television stations. And the guys who worked for her were her department heads. They were her board of directors and the only board of directors she had. C. K. (Elea-nor's nephew) was the editor and vice president. She had already made it clear he was to succeed her. I got along well with him from the start,

4. Interview with Jane Cote, San Jose, California, October 26, 2002.

and I got along well with her from the start. And I liked them both. They were simple and unpretentious. It was a totally different environment from Knight Ridder. Very low-key. To be ambitious was a no-no."

But Potts soon found himself in the same role that Chapman had assumed back at Knight Newspapers, making things work with more transparency and efficiency. He did it with a laid-back California style.

"One of the attractions of McClatchy to me was that it was private. For most of my early years at Knight, it was private. The public company pressures weren't unbearable but it sure changed the environment at Knight. And particularly when it became Knight Ridder, there was much more pressure for profitability, and it was a little less emphasis on personal relationships. Part of it was size. McClatchy was the opposite of all that. It was a smaller environment. It was very private. It was very personal.

"C. K. was a guy with a wide open mind. You could walk into his office and say, 'C. K., I think we ought to buy *The New York Times* and here's how we can do it.' If you made a credible case in twenty minutes, you'd be doing it. And that's not too much of an exaggeration. He didn't want an hour, just twenty minutes. I loved working that way. I had spent a year or two in Charlotte trying to get a new computer system approved by the corporate office in Miami and had gone through paperwork and jumped through all sorts of hoops and still hadn't gotten the damn thing. In McClatchy, all I had to do was walk in and say 'C. K., we need a new computer system.' He might call the financial guys and say do we have enough money to buy a computer system, but that was it.

"It was just a lot simpler and a lot easier to do things. Now if you wanted to do something like change a picture on a wall, that could be a major issue. One of the few times Eleanor ever got mad at me, or not even me but a guy who worked for me, wanted to change the wallpaper in the cafeteria. It had been in there for like fifty years. And we didn't think to ask Eleanor, and when that wallpaper was changed, it was a major blunder."

Al Neuharth

A perceived need to make operations more efficient was also what led Gannett Company to raid the Knights for executive material. Al

Neuharth was assistant managing editor of *The Miami Herald* when Jack Knight and Lee Hills decided to move him to what Knight called the "hardball" market of Detroit. Their *Detroit Free Press* faced a difficult competitive situation. The Knights had lost their bid to buy the failing *Detroit Times,* and *The Detroit News* snapped it up, acquiring enough circulation to be the dominant daily in Detroit.

Neuharth was just the kind of scrappy competitor they needed. Raised by a single mother in difficult financial circumstances in South Dakota, he served with the Eighty-sixth Infantry Division, part of George S. Patton's Third Army in Europe. After the war, he attended the University of South Dakota on the GI bill and took a fifty-dollar-a-week job with the Associated Press. Then he launched his first business venture, a statewide sports weekly called *SoDak Sports.* It failed, and Neuharth got as far away from South Dakota as he could, landing a reporting job at *The Miami Herald* in 1954. "Getting to the top," he would say later, "means taking one smart step at a time. Managing, maneuvering, manipulating your way from one stepping stone to the next."[5]

Gannett was a smaller outfit, but that step gave him more responsibility, as operating head and general manager of the two Rochester newspapers, the *Times-Union* and *Democrat and Chronicle.*

"We were too small, a little old shit-kicker company from Rochester, N.Y., to get a hell of a lot of attention," he recalled.[6] He got attention by drawing on his knowledge of Florida and persuading Gannett to invest close to $10 million in a new daily for the booming Cape Kennedy market. *Florida Today* was launched in 1966, and it broke into the black less than three years later.

Going Public

Newspaper companies go public for different reasons. For Gannett, it was to raise capital for acquisitions.

"We were a newspaper company and we wanted to grow as a newspaper company," Neuharth recalled. "But as a small, private company there was a limit to how much we could leverage ourselves." After a major

5. Al Neuharth, *Confessions of an S.O.B* (New York: Doubleday, 1989), 69.
6. Interview with Jane Cote and Philip Meyer, Miami, August 7, 2002.

acquisition, ". . . our borrowing capacity was pretty much at its limit and we couldn't see where we were going to go unless we had a public tank to dip in."

In 1967, when Neuharth was executive vice-president, Gannett got its public tank, and further acquisitions began. By 1979, the company owned seventy-eight daily newspapers, a national news service, seven television and fourteen radio stations, billboard operations in the United States and Canada, twenty-one weeklies, and the research firm of Louis Harris and Associates.[7]

Knight Newspapers went public two years later, but for different reasons. The Knight brothers were getting along in years, and they worried about the possibility of the company being broken up or sold to pay the estate taxes. Correspondence collected by Jack Knight's biographer, Charles Whited, shows what the older brother was thinking. In a letter to Basil "Stuffy" Walters, who had been his editor at the *Chicago Daily News,* Knight wrote: "Under our peculiar tax laws, which prevent the accumulation of 'excessive reserves,' a company which might wish to remain small has either to pay this money out in dividends or acquire another property. Thus, the small get bigger. Strange reasoning, isn't it?"[8]

In a letter to his brother Jim, he continued to worry about the temptations of growth. "I think we have objectives other than simply trying to see how big we can become. . . . Sometimes the lure of bigness tends to make newspaper publishers forget their prime responsibilities."[9]

After the public offering in 1969, the urge to grow was quick to kick in. The first big acquisition was Walter Annenberg's *Philadelphia Inquirer* and *Daily News.* The deal did not have the unanimous support of the executive committee. Jack Knight voted "no."

For McClatchy, the motivation for going public was more like Knight's than Gannett's. As Erwin Potts recalled the situation, the decision came as a surprise.

"I remember one day C. K. walks in my office and says 'I think we ought to go public.' And he knew how I felt about going public because

7. http://www.gannett.com/map/history.htm (retrieved October 4, 2003).

8. Charles Whited, *Knight: A Publisher in the Tumultuous Century* (New York: E. P. Dutton, 1988), 272.

9. Ibid., 273.

he knew that's one of the things I liked about McClatchy was that we weren't public. I looked at him like I thought he was crazy. What's this all about?

"He said the kids are getting older and we need to think about what happens if I die or if Jim McClatchy, his older brother, dies. We could have problems. I think that was one of the primary motivations. He was right. Two years later, he was dead. And the transition was fairly orderly whereas it would have been very chaotic if it was a privately held company.

". . . the shares were already on the market. There was liquidity. People were able to sell shares for estate tax purposes or whatever. It clearly had been the right thing to do, although I didn't think so at the time and sort of made a pest of myself over the issue. I did not make any bones about it. I told him I don't think we have to do this. I just don't think there's any compelling need to do it. But once he made up his mind about what he was going to do, he was a very strong guy." The company went public in 1988.

But C. K. McClatchy did take a precaution that the leaders of Knight and Gannett had not. He saw to it that two classes of stock were created. The B shares, a majority of the total, were kept in the family and given a 10–1 voting advantage over the A shares that went to the public. Over time, as various family members died or cashed out, B shares were converted to A, and the public's proportion grew, but the family still maintained an overwhelming majority of the voting power.[10]

Adjusting to Wall Street

Gannett Company was the first traditional newspaper company to embrace Wall Street, and the relationship was smooth. Neuharth got it off to a good start by realizing that analysts were graded on their short-term performance, and so he created a program of earnings management to give them the quarter-to-quarter predictability they wanted. He also educated them about the newspaper business.

"In those days, Wall Street knowledge of newspapers was limited to

10. By the time McClatchy's 2002 annual report was prepared, there were 19.5 million A shares and 26.5 million B shares.

New York City," Neuharth recalled in 2002. "And most of them were dying. So they were sure it was a dying industry. All they had to do was look around. You went from 10 newspapers to four in five or six years. And so they (investors) thought union problems, high cost, television dipping into your advertising pie . . .

"And so . . . we tried to educate them that the newspaper business was not New York City. The newspaper business was Rochester and Elmira and Westchester and New York state. And it was the small and medium sized cities around the country that represented the newspaper business, and they were all doing damn well. Some of them had licenses to steal, as you know."[11]

John Morton, a former journalist who became an analyst and then a newspaper consultant, has said that Wall Street analysts would rather see a company increase its earnings 400 percent in steady increments than gain 500 percent in zigs and zags.

"I know it's irrational," he told a seminar in Chapel Hill, "but it's exactly the way Wall Street thinks, and, of course, the person I was quoting was Kay Graham who very famously once said that . . . we don't manage our earnings, we'd rather get from zero to 5 in zigs and zags than to 4 in smooth lines. She told that to a bunch of security analysts from major financial institutions, and a whole bunch of them went out the next day and sold out their Washington Post stock . . .

"There were, of course, buyers on the other side of all those transactions. I once pointed out that somebody who was a buyer at what was then about $17 a share recently enjoyed a price of about $560 a share."[12]

Neuharth's level-performance strategy was made easier by the relatively small size of the newspapers in Gannett. A small-town monopoly newspaper owns the tollgate on the flow of information between a retail merchant and the customers. That's why he called owning the newspaper "a license to steal." Big-city papers had more chaotic markets and a greater likelihood of competition. They were also more likely to have to deal with unions. But earnings smoothing, as Neuharth recalled, was "a piece of cake" for monopoly newspapers in smaller and medium sized communities.

11. August 2002 interview.

12. John Morton, remarks to Seminar in Media Analysis, School of Journalism and Mass Communication, University of North Carolina at Chapel Hill, April 22, 2002.

"We managed it for quarterly earnings gained. We just shifted the cost. Nothing illegal. You just figure out what you're going to do, when you're going to build a new plant here or there or somewhere else and smooth it out. Managing monopoly newspapers is the easiest executive job in the world . . . particularly if they're in small or medium sized communities because you can manage everything. Newsprint is not as big a factor. You get into the big cities, the Canadian newsprint manufacturers have a hell of a lot to say with whether you have up or down quarters or years. It's much less of a factor with small newspapers."

There were some external factors that made it easy for Gannett and other newspaper companies in the 1970s. Cold-type printing technology was replacing the old hot-lead system by then, and computer pagination was moving work from the composing room, which was more likely to have a strong union, to the newsroom. In the early cold-type processes, stories and headlines were printed out on paper and pasted up on page forms whose images were transferred to the printing plate by a photographic process. That technology eventually evolved into a direct computer-to-plate method, which meant that copy editors in the newsroom could do all of the work formerly done by compositors in the back shop. Some of the cost savings were applied to better printing with more color, better detail in the photographs, and ink that didn't rub off on the reader's hands. That still left plenty for the bottom line.

Another factor that helped smooth earnings growth was inflation. One of the generally accepted accounting principles (GAAP) is the money-measurement principle. A newspaper might win Pulitzer prizes, have an appealing design, and serve as watchdog and guardian of its community, but all accountants look at is the nominal value of the money it makes. When there was double-digit inflation, that factor alone could account for most of the 15 percent annual increase in nominal earnings that Gannett used to impress Wall Street.

Accountants don't focus on nominal dollars out of ignorance. There's just no uniformly agreed-upon method for arriving at inflation-adjusted dollars. Both the Consumer Price Index and the Gross Domestic Product Deflator have applications for limited ranges of cases. Keeping accounts on an apples-to-apples basis for a wide range of business applications to compare at a given point in time is easier with nominal dollars.

By the 1980s, inflation began to ease, and it became harder to squeeze new savings from innovative production technology. It was also then that Gannett delivered a shock to Wall Street: a risky new venture, *USA TODAY*. Looking back years later, Neuharth saw it as proof that the company had not been deterred by Wall Street from making long-term and risky investments.

"If you want to look at the numbers, our stock went to hell when we announced *USA TODAY*. We were ridiculed. We were convinced it was a long-term investment that had a good chance of paying off. Not a sure bet, but a long-term investment that had a good chance of paying off pretty big. And if it didn't we could slash a good bit of what we were spending on our other newspapers. So we paid no attention to Wall Street. Had we, we wouldn't have launched it, (or) we would have folded our tent after the first year."

Gannett's early success with Wall Street did not go unnoticed by other newspaper companies. In my brief time at Knight Ridder corporate headquarters (1978–1981), some of us in middle management cursed Neuharth for setting the profitability bar so high. We had larger papers, and the price of newsprint was a major and uncontrollable factor. We also had chronic labor problems in Detroit and Philadelphia. But Alvah Chapman, who was present for the transition of Knight Newspapers from a private to a public company in 1969 and who oversaw the merger with Ridder Publications in 1974, remembered only admiration for Neuharth.

"I felt he was setting a standard to move us all up the scale," recalled Chapman. "That's all."

In Chapman's view, Wall Street's preoccupation with quarterly earnings growth was not all bad. "There's some discipline there that a well-managed business needs to be aware of at least to protect itself. . . . You program your costs to the extent that you can, some of it you can't program, you can't control. You have to go ahead and take them when you get them. To the extent you can manage your business so that you do no damage to the quarterly earnings, it's better to do it that way."

In Chapman's mind, the connection between profitability and quality journalism is obvious. He was asked if taking the company public inhibited it from making long-term investments in quality.

"I don't think so. We made decisions like merging Knight Ridder after we went public. We expanded our Washington bureau, doubled the size of the Washington bureau, after we went public. We added our overseas bureaus after we went public. We started Business Monday after we went public. We started the Neighbors sections after we were public."

He remembers the statistics from his administration of the company in detail from the merger (he oversaw the merger and became CEO two years later) to his retirement.

"In my 15 years, from 1974 to 1989, Knight Ridder's stock grew 23 percent, compound growth rate. We had 15 straight years of increased earnings per share. We won 37 Pulitzer Prizes in that period of time . . . Each year . . . we increased our contributions budget for community responsibility. So there's responsibility to the readers, the community, the employees. We were in the 100 Best Companies to Work for in America twice. The first two editions that came out, both editions covered Knight Ridder. That's not so now."

But the pressures of being a public company were felt at lower levels. I worked for James K. Batten when he had the painful job of carrying the bad news about earnings requirements out to the individual newspapers for which he was responsible. The emotional burden was heavy. From the news-side perspective, the main tool for smoothing earnings was the contingency budget. Under normal circumstances, a budget is a planning tool. Under the contingency system, editors had to produce several layers of planning with each budget. If revenues fell below a certain point, a contingency plan was triggered which meant, in effect, a budget cut in the middle of a planning year. There were no layoffs in those years, but projects got postponed and staffs were thinned through attrition in times when advertising was down.

It was in that period that Erwin Potts decided that managing a Knight Ridder newspaper wasn't fun anymore, and he jumped to Mc-Clatchy. When that company went public in 1989, he took advantage of the family voting power that made it possible to keep Wall Street at more of a distance. Some of his early Wall Street conversations contained echoes of John S. Knight two decades earlier.

"We said early on to the analysts, in any public presentation we made, that we consider ourselves a quality newspaper company, and we did not intend to operate in a manner that required quarter-to-quarter gains in profitability constantly. If people wanted that kind of invest-ment, they really shouldn't invest in us. We're a long-term good investment. And that's proven to be the case, if you look at our stock. We think that over a period of time, we'll be good. We'll be adding value to your investment as time goes on. But we're not going to do it like hamburger franchises. We're going to manage the business the way we think good newspaper people and good business managers should manage it, for long-term value and also because we think that a good newspaper has to be a good citizen and that those are not incompatible objectives."[13] As the company grew, Wall Street became more interested, Potts recalled.

"We went to the forums they had in New York City and we made presentations. We were attentive to the analyst community. We didn't ignore them. But we didn't feel like we had to dance to their tune if it was contrary to what we thought were the best interests of the company. And I think that's an advantage that we've enjoyed over the Gannetts and the Knight Ridders and the other companies that are not positioned the way we are. I don't want that to sound like the management is not responsive to the financial community. It is. After all, if people don't invest in the company outside the family, then the stock isn't going to have a public value and is not going to be worth anything to the family or to anybody else. You can't ignore those things. We've always had a profit motive, a strong profit motive."

Like Chapman, Potts saw some benefit in the discipline of the market.

"You've got to make money, in any business. In the newspaper business you've got to make it if you want to do the things you need to do to make newspapers better. Our company is a good example in the past of how those things could work against you if you didn't make money. *The Sacramento Bee's* building was falling down back in the 1970s. The presses were ancient. So was our pre-press equipment. We needed everything under the sun but we couldn't buy it. We'd have to go to the bank and beg.

"And today, we can invest in equipment if we need to, to make a better-looking newspaper. We can invest more in staff in newsrooms. And we continually reinvest in the newsroom. Gary Pruitt, my successor, is fond of saying this, and I think it's very true: as a company, historically, we have never been boom or bust. If the economy's good we don't go

13. October 2002 interview.

out and hire ten new reporters because the economy's good. When it goes bad, we don't lay off either.

"But we have steadily improved. We're always, constantly investing in improvement."

One of the benefits of this even-handed management is a happier workplace.

"People feel better about the company," said Potts. "They know they can make a career there. They know they can make a commitment to the company. The company has a commitment to them, as well. That's changed a lot in the business world at large. Double commitment back and forth doesn't seem to apply in many companies anymore. We've taken a lot of care with that over the years."

The Next Generation

Al Neuharth and Alvah Chapman both retired in 1989. Chapman's successor at Knight Ridder was James K. Batten, who had given up a job he loved, editor of *The Charlotte Observer,* to get on the corporate ladder. Buzz Merritt recalls visiting him in Miami. They were old tennis buddies, and Merritt had been Batten's first editor at *The Charlotte Observer.*

"He said let's go get a sandwich, and we got in his car, and I said, 'Jim, why in the world did you give up the editorship of a great newspaper, with all your news background, in order to become a corporate officer?' And his response was, 'Somebody has to watch the bad guys.' "

In the top job, Batten remained an able advocate for the news side, but an untimely brain tumor forced him to yield the reins to P. Anthony Ridder several years ahead of the planned schedule. Batten died in 1995.

Tony Ridder had earned the nickname "Darth Ridder" among the editors when he performed the task that Batten had hated, carrying the bad news from headquarters about budgets to the individual newspapers. Ridder had been a reporter early in his career, but his management jobs were on the business side. That made him a logical successor to Batten, given the Knight tradition of alternating between business experience and editing experience in the top job. What was lacking was a strong number two and heir apparent to protect his back on the news-editorial side. An anecdote from Buzz Merritt illustrates the problem.

"We had an editors-only meeting shortly after Jim died. . . . (Ridder) gave his usual speech about you can have quality and high returns, too. And then there was a traditional Q and A. This is a roomful of editors, and somebody, I forgot who it was, asked what worries you, what keeps you up at night?

"He thought for a minute and said, 'Electronic classifieds.' And the air just went out of the room. And the evening was over."

Rationally, Ridder's answer to the editor's question was a sound one. If the Internet becomes the substitute technology that destroys newspapers, the effect is likely to be felt first in classified advertising. But a better answer for that audience would have been phrased in terms of the danger to newsrooms in that very real threat. If Batten had been asked that question, he might have given the same answer, but couched it like this: "My fear is that electronic classified advertising, which is starting to grow, will become the leading edge of a force that deprives of us the means to carry out our traditional First Amendment responsibilities to our readers and their communities."

Tony Ridder's biggest problem was not of his making. It was on his watch that the game got much harder. The cost savings from new production technology were about used up, and newsprint prices took a big jump in 1995. In constant 2003 dollars, the eastern U.S. price went from $580 per metric ton in 1994 to $797 in 1995. That made newspaper managers think twice about trying to maintain circulation levels in the face of such a high variable cost.[14]

The price of newsprint eventually found its way back to the pre-1995 level, but it took the rest of the decade. Newspapers could still show earnings growth, but they had to raise prices and cut costs to get it done. The exuberant investing style that produced the stock market bubble of the 1990s turned the analysts of Wall Street into celebrities and intensified the tendency toward short-term thinking.[15]

It also raised the visibility of analysts who worked on the sell side. The analyst community is divided into the sellers of research, who generally work for brokers and investment bankers, and the buyers of research who represent large institutional investors such as pension funds,

14. Newspaper Association of America, *Facts About Newspapers 2003.* The constant-dollar adjustments are mine, based on the Consumer Price Index.

15. Landon Thomas, Jr., "Wall Street's Harsh New Reality," *The New York Times,* August 17, 2003.

insurance companies, and endowed nonprofit organizations. Pressure on public companies came from the sell side. John Morton, the former analyst, described the incentive structure:

"There is this great deal of pressure on the buy side to have their portfolio increase, because that's basically how they're graded, and so the sell-siders want to sell them information that makes them happy, and so they put pressure on the companies. I mean it's exactly like that. And that's where the pressure comes from . . . It's the pressure particularly from sell-side analysts to produce predictable events and no surprises.

"When I first got into this back in the early seventies and I was a Wall Street analyst, it was unheard of for a company to give earnings guidance to Wall Street. . . . But they all do that now, except for *The Washington Post*. . . . They basically tell Wall Street what they're going to earn, and, and if there's anything that happens to cloud that, you know, they'll alert you right away because they don't want there to be any surprises."

Short-term prediction, in Morton's view, became more important than fundamental research, and that's a reason he left the business.

"You suddenly had analysts who don't do any research at all. They're recommending companies out there that, if you looked at their cash flow balance, were heading for a cliff."

His view is echoed by people on the other side of the analyst-company relationship. While the men who were CEOs at the time newspapers started going public had fairly benign views of Wall Street, managers at the turn of the century were more guarded. Douglas H. McCorkindale, who took the reins at Gannett from John Curley in 2000, said, "The quality of analytical work on Wall Street, in my view, diminished in the late nineties. I think that too many of them didn't do their homework. I was fascinated by analysts issuing earnings estimates on Gannett and writing pieces about Gannett and never calling us and asking us a question . . ."

McClatchy's Gary Pruitt, who became CEO in 1996, took a parallel view. "I think the analysts don't do enough independent analytical work. What they do is too often just repeat what management has said as opposed to doing independent work."

This weakness creates the need for more intensive communication efforts on the part of the company, according to Pruitt. "Sometimes if you don't speak to the analysts or don't give guidance or don't talk about what you are doing, I fear we cast our fate to folks who are going to define the company for us. And I don't want to do that."

Both Pruitt and McCorkindale were lawyers who joined their respective firms as general counsels. McCorkindale was chief financial officer before taking the top job at Gannett, and Pruitt was vice president of operations and technology for McClatchy.

The analyst-company relationship was complicated by Wall Street corruption in the 1990s. The crash of the dot-com bubble brought attention to conflicts of interest among analysts whose pay structure rewarded them for bringing in investment banking business. This arrangement created a temptation to make a company look better to investors than it really was, in order to land it as a client.

Tony Ridder remembered that period. "Here's somebody that's really supposed to be like a reporter bringing, in effect, an ad salesman with him, saying, 'Look, I'm going to do the story on you and right next to me is an ad salesman. Why don't you spend more money in *The Kansas City Star?*' "

The same thing happened to McCorkindale at Gannett.

"I've had analysts come in here, with their investment bankers, and make the pitch as to why we should hire them and hint how positive their report could be. I told them to get the hell out of the office. That happened during the last couple of years."[16] Ridder saw reforms taking hold.

"The thing is," he said in 2002, "it's hard to generalize about analysts because most are very principled. I think 95 percent of them are damn smart. They ask very good questions. They might like to get more business somehow, but I think most are honest about it.

"I think some are interested in the quality of the journalism. Some are not. Some just view it as a business. Some don't. Some believe that there's a relationship between the quality and the future growth prospects."[17]

16. Interview with Philip Meyer in McLean, Virginia, and Jane Cote by telephone, June 10, 2003.
17. October 25 interview.

One additional source of stress for editors at both Knight Ridder and McClatchy in the nineties was the continuing movement to a more centralized management. Unlike editors at Gannett, they had become accustomed to a system of editors and business managers reporting separately to corporate headquarters. That meant every editor had an advocate at headquarters, but it bucked too many problems to the top. At McClatchy, Potts started to change that in 1991 when he named Gary Pruitt publisher in Fresno. Under the local-publisher system, one person reported to headquarters on behalf of both the news and business sides. There was efficiency in that because conflict got resolved at a lower level. But it left editors feeling more vulnerable to pressures from the business side.

When Merritt was editor in Wichita he saw one unexpected consequence of centralization. Papers started buying basic supplies and services such as travel and film through the corporation's central purchasing service. That saved money, he said, but the business managers "were astonished when the camera business in Wichita we bought our film from stopped advertising in the newspaper. They were stunned . . . Now how're you going to make up for the fact that the camera company isn't advertising in our newspaper? Publishers and owners and people who are bosses have to be convinced that the financial success of their newspaper is dependent on all those things. It is dependent upon the success of the community, not only financially but civically and every other way."

Layoffs also have an impact on the community. If the reporters are secure and happy, the community is probably going to be happier with their product. In 2001, a massive wave of reductions in force through buyouts, attrition, and layoffs swept through the industry. Knight Ridder's cuts drew more attention than most. One catalyst was the very public resignation of Jay Harris, the publisher in San Jose, and his emotional farewell to the American Society of Newspaper Editors.

Harris had gained recognition as an early adopter of precision journalism when he used a computer to do investigative reporting for Gannett's *Wilmington* (Delaware) *News Journal*. While on the faculty at Northwestern University, he designed ASNE's first census of minorities in newsrooms. After he rejoined Gannett as a national correspondent, Jim Batten recruited him and put him on Knight Ridder's executive ladder.

He was Tony Ridder's assistant before becoming publisher of the *San Jose Mercury News.*

His problem with the budget cuts in San Jose, he said, was not just their magnitude, but the rhetoric in the deliberations, which focused "myopically on the numbers."

"What troubled me, something that had never happened before in all my years in the company, was that little or no attention was paid to the consequences . . . There was virtually no discussion of the damage that would be done to the quality and aspirations of the *Mercury News* as a journalistic endeavor, or to its ability to fulfill its responsibilities to the community."[18]

In reacting to hard economic times, Tony Ridder was very mindful of his responsibility to shareholders as well as the behavior of his peers. Aggressive earnings management is built into investor expectations.

"I think they also expect we'd do what other public companies are doing. If the whole industry was basically going along with the flow, then it would be a lot easier to go with the flow. But when the industry is reducing head count and cutting back on spending, the price of our stock would be killed if we didn't react to the situation."

When Rick Edmonds analyzed the situation in 2002, he found that Knight Ridder had been reacting more strongly than other companies. The company's stated goal in response to the economic downturn that began toward the end of 2000 was to trim news staffs by 10 percent. At other public companies, in Edmonds' analysis, staff changes ranged from no more than 5 percent cuts at worst to modest increases at best.

"Knight Ridder accounted for an astonishing 60 percent of the net cuts among public-company newspapers in this analysis during 2001," Edmonds reported. Cutting staff size was nothing new at the company, but it had usually been done through attrition. Layoffs were quicker but came at a greater social and human cost.

There was an explanation for Tony Ridder's actions in the company's history. The news staff buildup under Chapman and Batten had made analysts uneasy because it departed from the only gauge they had, the industry average. "So," said Edmonds, "as management said, the 2001

18. Luncheon Address by Jay Harris, American Society of Newspaper Editors, April 6, 2001. Posted at http://www.asne.org/kiosk/archive/convention/2001/harris.htm (retrieved October 20, 2003).

cuts were more a remedial action to bring staffing levels closer to the industry norm than a pace-setting move."[19]

But what if the "norm" was the wrong strategy? What Jane Cote and Jerry Goodstein have called "herding behavior" among analysts can easily translate into herding behavior by newspaper companies. Few analysts ever want to stand out as different, because "joining the consensus provides cover" for their reputations.[20] Companies can seek the same cover.

Gannett went through the 2001 downturn without layoffs only because it was already at or below the norm. There was nowhere to cut. Its staffing had been bare-bones to start. McCorkindale saw some irony in this.

"There are a lot of good analysts who do a consistently good job, ask questions, blunt questions, many questions we don't want to answer. But one of the things they see is that Gannett doesn't have a lot to cut because we don't have a lot of extra cost to begin with. Therefore they focus on a company that's more casually managed, where they can cut and get a better result. They'll have that type of company on their buy list. And there's nothing I can do about that except to say, 'Don't you pay for management?' To be surprised . . . because all of a sudden advertising is down, doesn't make sense. If you're paying attention to this business, which is not that complicated, you will not be surprised. It's a pretty straightforward business. It's a very cash flow positive business. You know what the receivables are . . . I don't know how you can be surprised if you are paying attention."

McCorkindale did not believe that Gannett would behave any differently if it were a private company—and might even be at a disadvantage.

"A number of companies that have had family votes to protect the management structure were very comfortable. And, in many cases, they'd get into union agreements that were just silly. It had nothing to do with compensation. It was agreeing to work rules. They're all changed now. If they had been on the firing line to perform, they might not have been

19. Rick Edmonds, "Newsroom Staffing: Public Companies No Worse Than Private," Poynteronline, www.poynter.org. Posted December 5, 2002 (retrieved June 5, 2003).

20. Jane Cote and Jerry Goodstein, "A Breed Apart? Security Analysts and Herding Behavior." *Journal of Business Ethics* 18:3 (February 1999), 305–14.

quite so comfortable. So, no, I don't have any trouble, and we don't get any real financial pressure from Wall Street that we react to."

Some incidents in Gannett's history bear him out. The creation of *USA TODAY* was a prime example of a public company going against the short-term concerns of the analysts to make a large and risky investment. McCorkindale remembered a more recent case.

"We bought Multimedia (Cablevision, Inc.) in 1995. We stepped up and I decided we're going to buy the whole thing. Everybody wanted to buy a piece. I said 'No, we can do the whole thing.' Wall Street immediately came to the conclusion that Gannett knew nothing about the cable business, so we should sell it right away. Get rid of it for $800 million. But we sold it four years later for $2.7 billion. If we had paid attention to Wall Street, we would have lost $2 billion for the Gannett shareholders.

"You can't let them dictate your management. You can listen to them. You can try to keep them happy. But, at some point, you're going to disappoint them. And we're going to tell them, 'No, there are no more bodies to be cut out of Gannett. So, if you don't like that, go find a company that's poorly run and then you'll get some bodies cut out.' I've been doing this for thirty years, and I don't have any problem with just telling them that they're off base."

On the other hand, Gary Pruitt, who replaced Erwin Potts as CEO of McClatchy in 1996, believes that family control carries a strong advantage. McClatchy, like Gannett, avoided layoffs in the 2001 downturn, but for a somewhat different set of reasons.

"While the newspaper industry is cyclical, it is not so cyclical that in a downturn you must have layoffs," he said. "Looking at it historically in the United States, downturns are relatively brief, typically eleven months or so. And growth is much more prolonged. Usually five or more years, often longer. We're very fortunate to have more up years than down years and if you can run your business consistently, you are better off. So, we try to remain consistent in good times and bad . . .

"In most cases, I think layoffs say more about how the company was managed than they do about the economy. That's a quote that's gotten me into trouble. But I do think there is truth to that. It felt very late cycle to me in the late 1990s and we did not needlessly bulk up so that we were not in an untenable position when the downturn occurred. Turnover is such in the American economy that you can accomplish a lot through attrition.

"Also, on a human level, I really felt for anyone in a tough economy. But for our employees, it sounds old fashioned, but we really do have a commitment there and a compact. And we're not going to have layoffs if we can avoid it."

Pruitt was mindful of the competition from new technology and believed that this competition, or its potential, created pressure to improve quality through good times and bad.

"We always say to our papers, your challenge is no matter what, the paper must improve. It always drove me crazy . . . where in a downturn, news hole cuts were made and the paper got worse. . . . And I always thought restaurants don't make food worse in a downturn. Car companies don't make cars less safe in a recession. Clothing companies aren't making lower-quality clothes. If they did, we wouldn't be their customer. We would resent them for it, and we should resent them for it. And I wouldn't give them business even in good times after that. Why is it OK to make a newspaper worse in a recession? That's your excuse for making your product worse? It makes no sense. We thought what we need to do, even in a downturn, is plan how we will improve each paper."

It is easier to say this, of course, if family control makes your newspaper immune from a hostile takeover. Analyst Lauren Rich Fine observed that there never has been a hostile takeover of a newspaper company, and said she believes there probably won't be "given the gentleman-like nature" of the publishing fraternity."[21]

Pruitt agreed, but with a slightly different take. "I think one of the things that constrains them is they think if they take over someone in a hostile way, they will be left out of future opportunities that they may want. I think there's a practical side to it that has nothing to do with being a gentleman."

However, there are potential investors outside the gentleman's club of publishing who would not be so constrained, and Edmonds has suggested that there have been private equity firms looking for ways to take over Knight Ridder.[22] As the arithmetic in chapter 2 showed, there was a time when it would have been at least theoretically possible to

21. Lauren Rich Fine, interview with Philip Meyer, New Orleans, April 9, 2003.
22. Rick Edmonds, "Who Owns Public Newspaper Companies and What Do They Want?" Poynteronline, http://www.poynter.org. Posted December 11, 2002 (retrieved June 5, 2003).

pay for a company like Knight Ridder by selling off a surprisingly small number of its newspapers.

"Those private equity firms, it is a viable possibility," said Pruitt. "I would not rest as easily at night if we didn't have this two-class stock structure. And there may be actions that we're able to take now and rest comfortably that we're taking a long-term view. I don't know about day-to-day operations. I'm not sure they would change that much. But I think you would always be much more aware of the prospect that you may be taken over."

The Analysts

The relationship between analysts and the managers of newspaper companies is more complicated than it seems on the surface. For one thing, analysts can't tell management directly what they want it to do. That would make the analysts too much like insiders and run afoul of the Securities and Exchange Commission's full-disclosure rules. But they can convey their wishes and expectations by asking questions. And, since most of the questions clearly represent efforts to get a better estimate on the next quarter's results, management gets the point.

But Merrill Lynch's Lauren Rich Fine said the analysts don't really dictate management behavior.

"The companies that do things for us are wrong. Not one of us respects a company that is doing things for us. We'd rather them run their company, help us understand how they're running it, then we can decide how much their stock is worth."

The analysts are a little bit tired of journalists complaining about the discipline of Wall Street—especially in those cases where going public was a move to keep the company intact after the death of a founder.

"The real genesis of all the problems here," said Fine, "is newspaper companies went public for the wrong reason. They didn't care about the shareholders, they did it for their purposes. My frequent point that I've made is that if you don't like what it means to be a public company, you don't need to be public because, here's the good news, as many of you have pointed out, these companies generate a lot of free cash. Not one of them, not one needs to be public. So, that's where the real tension comes in between Wall Street and the public newspaper companies, which is, 'Hey, I didn't tell you to go public, but as long as you did, I expect a return.' "

William Drewry of Credit Suisse First Boston had a similar view. "If you want to be a public company, by default you hold yourself up to the scrutiny of the market. And the market demands growth. It demands return on its invested capital because the market buys the stock, is investing capital, and wants a return on it. There's a threshold that has to be crossed to have success."

And yet, the analysts have a surprising respect for those public companies that organized themselves to have some independence from their investors. Fine liked both the *Washington Post* and McClatchy for their long-term views.

"They went public for the exact more or less same reason (as Knight Ridder), to keep a franchise together, and they run it their own way. They are running for asset value over time. Their margins, I don't know off the top of my head what they are for the *Washington Post,* but they've got be about the lowest that we've probably ever seen. They couldn't care less about what I think about the margin. But they do all the right things in terms of investing, having a quality newspaper, and that stock has done magnificently."

Conversely, analysts don't like companies that always yield to Wall Street's shifting winds. Fine mentioned a firm, not a newspaper company, that "has a CEO that we love to refer to as somebody who ages but doesn't mature. He does things purely based on what he thinks Wall Street wants them to do. He buys and sells assets on that basis. As a result, everybody hates this company because they're afraid that the last person in the door gave him bad advice."

In contrast, Fine said, Wall Street appreciates a company that has a plan and follows it. She gives McClatchy credit for having a visible and consistent strategy. "They have been so remarkably innovative, yet they're everything we're supposed to hate, which is family controlled and caring about quality. And yet, look at how fabulously their stock has done."

Some analysts were critical of McClatchy's acquisition of the *Star Tribune* and thought the company paid too much for it. Maybe the company does overpay, said Fine, "but in the end, there have been great returns on these investments because they've ended up being smarter than we were."

From Gary Pruitt's point of view, it's more a matter of thinking farther ahead. He remembers the Minneapolis deal.

"When we bought the *Star Tribune* in Minneapolis in 1998, we actually paid more than our entire market cap for that company. It was quite dilutive. And I met with shareholders. Some of the shareholders were quite upset and other shareholders actually viewed it as a buying opportunity to buy more stock on the downturn. And the difference wasn't because they have different numbers. Everyone had the same numbers."

The people who bought on the downside came out winners.

"It has proven to be a great deal. It had all to do with time horizon. . . . That deal was done in March 1998 . . . In the five years from 1998 through the end of the 2002, I know this because we just had to put it in our proxy, McClatchy outperformed every newspaper company in stock performance."

Different analysts have different strategies and different time horizons. The sell-side analysts such as Fine and Drewry tend to have shorter outlooks than those on the buy side. But even among them, there is variance.

Merrill Lynch used to provide split ratings of companies—long-term and short-term outlooks. But that confused people, so they switched to doing just a single rating, and it tends to tilt toward the short term.

"There are times where I know a stock is going down, and I will have to tell people to sell because I know the stock is going down," said Fine. "But, where I liked our split ratings was I could tell somebody, short term, you should sell. But long term, a year from now, the stock is going to be higher. So if you want to suffer the transaction costs, if you want to have a stomach ache, just understand that it's going down but I believe it's coming back up again."

If newspaper companies can find a few quality indicators that could be measured and published periodically, would Wall Street care or even pay attention? Fine thinks so, at least in markets with print competition.

"You can't please every investor every day. You can hope that people see the virtue of how you're building the value of your company over time, which is all that you all are trying to get at, is to build the value, you build the quality, you build the circulation, and you're building revenue. You're looking at it that way. So you're trying to attract the value investor, who says I really see what you're doing. I'm going to make money with you if I invest over time, and I recognize those times where you're not going to look as good as somebody else. But a few years out you'll look better."

Those are the kinds of investors McClatchy is after. "We had, to some degree, tried to shape the type of shareholders we had, and have by defining the company differently and not wanting to have certain types of shareholders that would only be upset by our actions," said Pruitt. "So we generally have shareholders that have stuck with us a long time."

Analysts are easier to convince about the importance of quality journalism when a market is competitive. But Fine is typical when she expresses skepticism about the value of quality where a newspaper already dominates its market. "If I was the average person, not somebody working on Wall Street . . . and I lived in Cleveland, I'd be reading the *Plain Dealer.* It could be a great paper. It could be a horrible paper. I either want a paper or I don't. That's the thing I always have trouble wrestling with."

Bill Drewry has made money for his newspaper investors by realizing that the decline in circulation would come more slowly than many expected. His bullishness on newspapers has come "not so much from any belief in any great growth opportunities that newspapers have versus other media, but more of a belief that change would come slower than most people thought. I think that's been the common thread on Wall Street for a long time. That there's this major change dynamic that was going to just take hold of the newspaper sector, whether it was the Internet moving big chunks of the revenue away or the readership patterns shifting rapidly to more electronic. And I realize now that my core belief was that the change would not occur as rapidly, and I think that's what's happened."[23]

Newspaper reporters have a cynical expression that helps them cope with the inherent frustrations of their profession: "Good things happen in spite of management." For the best of the top managers, a parallel idea might be, "Good things happen in spite of Wall Street." Al Neuharth's tenacity in creating *USA TODAY* and seeing it through its ten years of red ink, Gary Pruitt's bold acquisition of the *Star Tribune* illustrate the value of managerial autonomy to get results beyond the limited vision of most analysts.

23. William Drewry, remarks at Poynter Institute, February 10, 2003.

In the spring of 1981, Lee Hills made his last address to the Knight Ridder shareholders. As an employee-shareholder, I was there. Lee had stepped down as chairman of the board two years earlier but continued some involvement in day-to-day operations. The company had been public for twelve years, and he was well aware of the tensions. Profits are necessary, he said, but, "One priority clearly stands above orderly profit growth and all other requirements. That is the long-range health of our company.

"It isn't always easy as a public company in America today—in *any* industry—to keep this top priority in mind. Many big traders in the stock market and their advisors decide to buy or sell depending on how this quarter's results compare with the previous quarter and the last year.

"I happen to think that this short-term pressure is unhealthy for American business . . . the short-term pressures must be resisted—no matter how beguiling—in favor of long-term success. We are not building a company that measures its life or its progress in ninety-day spans . . . we are committed to the idea that quality coincides with the interests of the shareholders."[24]

It was an echo of John S. Knight's opening remarks to the analysts twelve years earlier. It would always be necessary to do some things that they did not understand. It was necessary to not become their prisoner.

Jack Knight died in the last days of that spring. By coincidence, I had told him good-bye a few weeks earlier. The two of us shared a secretary in *The Miami Herald* building, and she alerted me when he was ready to make his annual escape to Akron ahead of the subtropical summer. My resignation from the company had been announced, and I went down to the office of the editor emeritus just off the newsroom.

He steered our last conversation toward quality issues. He didn't like the *Herald*'s latest design, which included a lot of attention-demanding black borders. The eighty-six-year-old philosopher-king of newspaper publishing waved the paper in the air, complained about the small type, and chafed at its "funereal look." We shook hands, I patted him on the shoulder, and we wished each other well. As I returned to my office on the sixth floor, I wondered what the next page of the history book would show.

24. Lee Hills, "A Commitment to Quality," Knight-Ridder Newspapers First Quarter Report, 1981, 9.

11

Saving Journalism

H O W F A R have we come? The material in this book is based partly on reporting and partly on sitting and thinking. So far, the reporting part has produced evidence for the following things to think about:

1. Newspapers that operate in places where they are trusted do better than newspapers in other places. (chapter 1)

2. Newspapers make money by owning the tollgate through which information passes between retailers and their customers. It's no longer an exclusive tollgate. (chapter 2)

3. The advertising market is aiming at ever narrower audiences. In the new competition, a trusted medium would still have an advantage. (chapter 3)

4. Influence is hard to measure. Community affiliation and trust are important components. (chapter 4)

5. Accurate reporting leads to more trust. Part of this effect is direct, but most of it is mediated through people with firsthand knowledge of the reported events. (chapter 5)

6. Newspapers that are easy to read have better home count-penetration than newspapers that are hard to read—but it erodes at the same rate. (chapter 6)

7. Shifting the form and substance of content, within the capacity available to most editors, doesn't make much difference. (chapter 7)

8. The public is very forgiving about errors in spelling and grammar—unless the spelling error is in the name of the person being asked. (chapter 8)

9. Well-staffed newspapers do better than those that are thinly staffed. (chapter 9)

10. Wall Street will support quality in journalism where there is competition. It is slow to recognize new media forms as journalistic competition. (chapter 10)

All of this is potentially useful information as we try to construct a new business model for news. And we do need a new model. Those of us who worked for newspapers in better times often wish we could go back to the golden age of newspapers. But it's over. The world moved on while we were thinking about other things.[1]

When newspapers performed consistent and visible public service—and many did in the golden age—it was because of the conscious choice of the philosopher-kings of publishing. These men and women had a long-range view toward two goals. The first was to dominate their respective markets. To do that meant providing more quality than their competition. In a fine confirmation of capitalist theory, the good really did drive out the bad and the less good.

Some mopping-up operations in the struggle for market domination were still under way in the first decade of the twenty-first century. The most interesting competitive situations were at the margins of major cities where they shared a metropolitan area defined, as the Bureau of the Census does, by strong internal social and economic links around an urban core. Dallas–Fort Worth, Miami–Fort Lauderdale, Raleigh–Durham, and Minneapolis–St. Paul are good examples. Newspapers in these pairs had historically maintained an uneasy truce, haunted by the aphorism from Harvard Professor Steve Star: "In the long run, it will not pay you to be in a market that you do not dominate."[2] Throughout

---

1. "The world moved on." Stephen King fans will recognize this phrase as the fate motif from his Dark Tower series. I like it for its dour sense of inevitability.

2. Steve Star was a Harvard Business School professor who ran a series of marketing seminars for newspaper executives in the 1970s.

history, from the medieval bazaar to Main Street to the giant malls, buyers and sellers of goods have found it efficient to converge on a central public sphere to do their trading and their socializing. In the twentieth century, the market's dominant newspaper was the glue that held that sphere together.

The battle to gain and keep that dominance made quality easy to justify, as Rick Edmonds established when he identified Knight Ridder's *Fort Worth Star-Telegram* as "the best-staffed newspaper in America." His case study showed how investment in quality held off the raiders from Dallas and paid off at the bottom line.[3] (See chapter 9.)

Al Neuharth has cited another example. In this one, it was his company that lost out to a better-quality product. It happened after he retired, but he still took it personally. Gannett went up against the locally owned *Democrat-Gazette* in Little Rock, and lost out to a local owner.

"Walter Hussman beat our tail by putting news in the paper and by having the community and the leaders of the community, including the advertising community, believe in what he was doing," said Neuharth. "Dillard's department store, headquartered in Little Rock, they believed in his credibility so much they pulled all their ads out of our Gannett newspaper. Big hit, big hit."

Some time afterward, Neuharth took a bus trip across the country with four of his oldest children. They stopped in Pine Bluff.

"The morning we left Pine Bluff I picked up the *Arkansas Gazette,* the northwest Arkansas edition. I read the goddamned thing for three hours on the bus. We got within a half hour of Little Rock and I called Walter, and I said, 'Walter, I'm pissed off at you. I spent my whole morning reading your damn newspaper.' He said, 'Well, I'm glad to hear that.' He's such a soft-spoken southern gentleman.

"He's still doing it. He's got a monopoly now . . . He puts news in the paper, a lot of it. He puts news that the people of northwestern Arkansas and central Arkansas want in the various editions and he's got a lock on the communities, he's got a lock on the advertisers, and he chased us out of town."[4]

3. Rick Edmonds, "The Best-Staffed Newspaper in America," Poynteronline, www .poynter.org. Posted December 6, 2002 (retrieved October 23, 2003).

4. Al Neuharth, interview with Philip Meyer and Jane Cote, Miami, August 7, 2002.

A Personal Decision

Once a newspaper owner gains domination of a market, however, the motivation for quality becomes less straightforward. The second source of motivation toward high purpose in the golden age was personal or family pride, and it was not universal. A private owner had enough wealth to engage in personal pleasures, and some found personal pleasure in producing influential newspapers. That personal decision on the part of some publishers—those that I call the philosopher-kings of modern media—was the main source of the golden-age legend.

Jim McClatchy expressed such a personal sense of mission when he said his family's newspapers were pitted against "the exploiters—the financial, political, and business powers whose goal was to deny the ordinary family their dreams and needs in order to divert to themselves a disproportionate share of the productive wealth of the country."[5] John S. Knight showed where his heart was by keeping the title of editor or editor emeritus to the very end. "There is no higher or better title than editor," he said. This is the publisher who wrote anti–Vietnam War editorials starting in 1954 when the French gave up on Indochina, and he opposed that war until it was over. Katharine Graham's support of her editors and reporters who uncovered the Watergate crimes was not motivated by profit but by her sense of civic duty. But being profitable gave her the luxury of being able to perform that duty, although she was modest about it. "By the time the story had grown to the point where the size of it dawned on us," she said, "we had already waded deeply into its stream. Once I found myself in the deepest water in the middle of the current, there was no going back."[6]

The golden age was different for different organizations, depending on who was in charge, but it generally fell into the period after World War II when the social mood began swinging away from conformity and national unity at all costs and began the long drift toward libertarian individualism. (Future historians might decide that the pendulum turned again after the September 11, 2001, destruction of the World Trade Center.)

Quality journalism became easier to support around 1970 when pub-

5. Quoted by Susan Paterno, "Is McClatchy Different?" *American Journalism Review* 25:6 (August/September 2003), 28.

6. Katharine Graham, *Personal History* (New York: Vintage Books, 1998), 505.

lishers first started paying serious attention to the decline in readership and decided to invest more. But its real driver was the will of the individuals who wanted more than economic return from their investments. Like the low-paid reporters who worked for them, they were in it for psychic reward, for a sense that they were making the world better.

There were even investors who appreciated the psychic reward. After Watergate, Katharine Graham received a letter from her newest and largest outside investor, the sage of Omaha, Warren Buffett.

"The stock is dramatically undervalued relative to the intrinsic worth of its constituent properties, although that is true of many securities in today's markets," he said. "But, the twin attraction to the undervaluation is an enterprise that has become synonymous for quality in communications. How much more satisfying it is going to be to watch an investment in the Post grow over the years than it would be to own stock in some garden variety company which, though being cheap, had no sense of purpose."[7]

It is hard to forget publishing's philosopher-kings. If only the present greedy proprietors would retire from the scene, we are tempted to argue, if they would allow themselves to replaced by more public-spirited individuals, the good times would come back. But that would be like Waiting for Lefty or Waiting for Godot. (Those guys never showed.) Journalism's new saviors will be different.

Those who would preserve the best of journalism's traditions should start with the premise that it is a business. The old convention that editors should be protected from all knowledge of the business side is hopelessly out of date. McClatchy, Knight, and Graham are proof that the two responsibilities can exist in one person. While the Watergate story was unfolding, Katharine Graham was mindful of her responsibility to shareholders and she was "frightened for the future of the Washington Post Company," but she took the socially responsible action.[8]

The Community as Market

People of my ilk with a news-editorial background are uncomfortable with the rhetoric of the business side. The folks on that side of the

7. Warren Buffet, quoted in ibid., 512.
8. Ibid., 507.

wall call the community a "market," reminding us of functions that
we'd rather not know about. But in fact, the concept of market is insep-
arable from the concept of community. As the Bureau of the Census re-
alizes when it defines metropolitan statistical areas, both social and
economic ties hold it together. The historic importance of newspapers
is based on facilitating both kinds of ties to build and maintain a public
sphere.[9]

Too often, editors believe that their ignorance about the business side
protects them. From my own years in the ranks, I am convinced that
the wall of separation is just as often used to limit the power of editors.
Once at *The Miami Herald,* Al Neuharth demonstrated how an editor
can punch a hole in that wall to strike a blow for news and community
service.

It was the custom, when Al was assistant managing editor in the
1950s, to make room for a major story by adding some open pages to
the paper—if the story was big enough and Jim Knight approved. Being
a prudent businessman, Jim always asked how much the additional
pages would cost.

Neuharth noticed that the business manager based the estimate on a
year's total production expense divided by number of pages. That
yielded the average cost per page. But that was a mistake because that
sum included each page's share of the fixed costs—depreciation on the
press, taxes, salaries, all the things that don't depend on the number of
pages in the paper on a given day. The relevant cost, Neuharth realized,
was the *incremental cost* of the additional pages, that is, what the *Herald*
would spend if the pages were added minus what it would spend if
they were not. In other words, just the variable costs—newsprint and
ink to create the additional pages plus any overtime needed to set the
type—should be considered. Al made that case, Jim Knight bought it,
and boosting the size of the paper got easier.

The reason newspapers were as good as they were in the golden age
was not because of the wall between church and state. It was because
the decision-making needed to resolve the profit-service conflict was

9. I use the term more loosely and less critically than Habermas when he said, "The
world fashioned by mass media is a public sphere in name only." See Jurgen Habermas,
The Structural Transformation of the Public Sphere, translated by Thomas Burger
(Cambridge: MIT Press, 1991), 171.

made by a public-spirited individual who had control of both sides of the wall and who was rich and confident enough to do what he or she pleased.

In today's world, most leaders of the press do not have that kind of functional autonomy.

We should not overromanticize the philosopher-kings and think that they never worried about money. Jack and Jim Knight knew how to hold on to a nickel. On my first trip to the men's room at the *Herald*, July 1, 1958, I washed my hands and was dumfounded to find that there were no paper towels, just a bin with leftover newsprint cut from the ends of the rolls into towel-size portions. I had seen the same trick at Faye Seaton's *Manhattan* (Kansas) *Mercury-Chronicle* in 1952 and mistakenly assumed the Knights would be higher-class.

So the pressure to downhold expenses (as the wire services would say when telegrams were charged by the word) is nothing new. And a few basic laws of money and its management have not changed. A news organization has to deal with financial constraints at three levels:

Level 1: It is necessary to earn at least as much as you spend. Even the *St. Petersburg Times*, which is owned by a nonprofit educational institution, has to pay for the ink and newsprint that it consumes, and it must put something aside to buy a new press when the old one becomes obsolete or wears out. To keep from spending beyond its means, any newspaper, no matter how idealistic and public spirited, needs a budget. To budget is to plan.

Level 2: The owner, whether an individual or a shareholder, is perfectly reasonable to expect that the return on investment will be at least as much as whatever the bank on the corner pays for certificates of deposit. In a way, this is a cost, like newsprint. Economists call it "opportunity cost."

Level 3: Investors in business enterprises are generally risk takers who need more than a steady income as compensation for the risk. They want the value of the investment to grow. This is especially true for shareholders who have many companies in many lines of business from which to choose. Value can be increased in several ways: making the product better and charging more for it; finding new customers; locating undervalued properties and making them more productive.

Going public forces a company toward level 3 in a hurry, if it wasn't already there. The McClatchy newspapers were close to level 1 when the company hired Erwin Potts. Everyone was content to chug comfortably in place so long as the bills and the salaries were being paid. The Knight newspapers were casually managed before Alvah Chapman started walking around with his clipboard.

Public ownership's imperative to grow the business is one of the things that has changed newspapers forever, but it is by no means the only thing. The other factor, and the most compelling reason to stop waiting for the philosopher-kings to come back, is the substitution of disruptive new technology for the printing press.

The philosopher-publishers could afford to be public spirited because of their monopoly positions. Like the Savoy family at Chillon, they owned a toll road. Theirs was in the form of a newspaper printing press, which is a very large, complicated, and expensive machine that breaks easily. It does not make economic sense to have more than one per market.

This constraint has not existed for every other truly new mass communication technology to follow the printing press. Let me explain with a more detailed review of the arithmetic of manufacturing that Al Neuharth used on Jim Knight.

When you sell an item, you set its price in the hope of recovering two kinds of cost, fixed and variable. Variable costs are those that vary in direct proportion to the number of items produced. If you produce 120 newspaper pages, you use twice as much ink and twice as much newsprint than if you produced only 60 pages. So ink and paper are variable costs.

Fixed costs are expressed, not in numbers produced and sold, but in units of time. Depreciation on the plant, taxes, and payroll are constant, within limited ranges, over a period of time regardless of the number of units produced. You can make and sell more or less than originally planned, but those costs are sunk and will not change.

The price of what you sell is set in an attempt to capture two elements: recovery of the variable cost plus a contribution to fixed cost and profit. Looking at it another way, contribution is what you have left after variable cost has been backed out of the price:

$$C = P - V$$

where C is contribution, P is price per unit, and V is variable cost per unit.

If you are in a business that makes things and sells them, you want to know your break-even point. It is calculated with the following formula:

BE = F/C

where F is fixed cost and C is unit contribution. The formula yields the number of units you need to produce and sell to recover your fixed costs. That's where you break even. Every additional unit will yield profit.

Let's say you are going to build a highly automated pencil factory. Total cost of materials and shipping—the only variable costs—amounts to three cents a pencil. You plan to sell the pencils to a wholesaler for seven cents, so the contribution to fixed cost and profit is four cents per unit.

Amortizing the plant and equipment plus the salary of the lone operator costs $200,000 a year. So the number of pencils you need to sell, if you are to break even the first year, is 200,000 divided by .03 or 5 million pencils.

The newspaper business is more complicated because it sells two things: the newspaper to readers—generally without recovering all of the variable cost—and space to advertisers, which provides enough contribution to fixed costs and profit to make it all worthwhile. But the basic distinction between fixed and variable costs is the same.

In the typical newspaper described in chapter 2, the variable costs of newsprint, ink, and distribution were 25 percent of the total cost. A newspaper cannot grow without increasing those. A broadcast station, radio or TV, in contrast, can keep pumping out the same signal to more and more people in a growing market without increasing either production or distribution costs. That was an amazingly nice business to be in during the early days when limited space on the airwaves made broadcasting a quasi-monopoly. Now that cable has taken the lid off the number of available channels, the business is no longer amazing, but it's still pretty good.

Bring on the Internet. Even though its combination of words and pictures looks like print on the screen, even though it can give you a paper product in your hand if you want one, it is more like broadcasting than print. The publisher has, essentially, no variable costs. If the customer wants the information on ink and paper, he or she provides them.

Newspapers have stopped growing, and not just because the public has gotten tired of them and prefers the new electronic substitutes. Newspapers already had natural limits of growth imposed by the structure of high variable costs. The new competition is free of that constraint and has a great deal of financial flexibility as a result. As the new century dawned, it was a sleeping giant poised to take over more and more of the traditional newspaper functions.

So why would anyone invest in a newspaper company today?

The Greater-Fool Theory

Remember the harvesting strategy brought up in chapter 1. Newspapers are not harvesting in the sense intended by Professor Porter because the industry is not yet in a clear end game. It has a track record of finding ways to cut cost, and investors hope that trend will continue. Analyst Bill Drewry (chapter 10) is in that camp. He made money for his short-term investors by foreseeing that newspapers would fade far less quickly than other analysts expected. This strategy sounds like a variation of the greater-fool theory. (I might be a fool for paying this much for shares in a fading industry, but I do so in the hope of finding an even greater fool who will pay still more for them.)

In this situation, it could be a mistake for newspaper advocates to urge longer-term thinking by analysts and investors. Their gloomy long-term view of the industry would argue against investing at all. Consider this excerpt from Lauren Rich Fine's May 2003 report:

> Long term, in our view, newspapers are likely to moderate their circulation volume declines but are unlikely to eliminate them altogether due to lifestyle changes and competition. We also continue to project ad market share declines . . . valuations seem to ignore the long-term secular declines, and are instead supported by deregulation speculation and a perceived cyclical recovery.[10]

10. Lauren Rich Fine, "The Newspaper Industry: Circulation Stalling but Efforts Continue to Improve Readership," In-Depth Report, Merrill Lynch Global Securities Research and Economics Group, May 29, 2003.

In other words, short-term factors are all that the newspaper industry has going for it. To save the traditions and the public service aspects of advertiser-supported journalism, it might be necessary to look beyond newspapers. We might have to look to the dot-com world or some hybrid. What could excite investors and be socially responsible, too? Analyst Drewry thinks newspaper companies might be the ones to do it. Their essential functions will always find a market, and necessity will make them innovators.[11]

It's a tall order. With new products especially, economic decisions can go far astray from real value, a problem observed in 1905 by Thorstein Veblen. Although he is best known for *The Theory of the Leisure Class,* he also wrote about businessmen who make money without making goods in *The Theory of Business Enterprise.* He saw how the financial system was creating conflicts for managers between their own short-term advantage and long-term service to society and to their own companies. If the managers "are shrewd businessmen, as they commonly are," he said, "they will aim to manage the affairs of the concern with a view to advantageous purchase and sale of its capital rather than with a view to the future prosperity of the concern or to the continued advantageous sale of the output of goods or services produced by the industrial use of this capital." [12]

Veblen was speaking of intangibles purchased only for the purpose of resale, not sale to an end user. Advantageous purchase and sale of goods that are headed toward an end user can add efficiency to an economy because each transaction adds value in some way. My grandfather Jacob Meyer was an excellent judge of cattle, and he supplemented his family farm income by spotting promising calves and buying them to be resold after some feeding and maturity made their value apparent to less skilled eyes. Saving them from premature slaughter created value.

But where capital is the good being bought and sold, value is, in Veblen's words, "in great measure a question of folk psychology rather than material fact." It depends "on the indeterminable, largely instinctive, shifting movements of public sentiment and apprehension."[13]

We would feel better about newspaper company shares being subjected

11. William Drewry, telephone conversation, December 18, 2003.
12. Thorstein Veblen, *The Theory of Business Enterprise* (New York: Charles Scribner's Sons, 1904, reprt., 1923), 157.
13. Ibid., 149.

to this kind of arbitrage if we could see it as a link in a chain of causation leading to better quality. But instead it pushes the other way. Wall Street fell in love with newspaper companies for their ability to cut costs and raise prices, and now that all the easy ways to do that are exhausted, there is not much left to do but cut quality. Jack Knight saw it coming in 1978: "And the rates are higher, all the time, and the type is smaller. So what you're doing is that you're charging the reader more and giving him less. Is that smart merchandising? I don't think so."[14]

Investors, as Lauren Rich Fine affirmed in chapter 10, can see the value of quality when two newspapers are competing. But what about the value of quality when newspapers as a whole are competing with a substitute technology?

Those of us who want to find economic arguments in favor of quality are going to have to face that issue. As the research for this volume has shown, and as a long string of research projects going back to the 1960s confirms, there is a positive correlation between quality and business success. But there is very little to suggest that quality is the prime cause rather than an incidental effect of profitability, except in those cases where two or more newspapers are contesting for market dominance.

The exceptions are the cases at the extreme low end of quality. The quality of most daily newspapers in the United States varies within a very narrow range. If there were more variance, we might see more dramatic effects.

An Extreme Case

Anecdotal evidence for this proposition can be found in the case of the Thomson Newspapers, the subject of a 1998 case study by Stephen Lacy and Hugh J. Martin. They chose Thomson because of its terrible reputation. C. K. McClatchy, in his 1988 Press-Enterprise lecture in Riverside, California, named it as one of three groups that produced "the worst newspapers in America . . . good newspapers are almost al-

---

14. John S. Knight, interview with Dan Neuharth, October 1978. From the transcript published as a memorial supplement by the *Akron Beacon Journal,* June 1981.

ways run by good newspaper people; they are almost never run by good bankers or good accountants."[15] And one of the Thomson executives had been quoted by a reporter for the group's own *Toronto Globe and Mail* as admitting that the company's emphasis on profits in the 1980s resulted in "cruddy" newspapers.[16]

Lacy and Martin tracked the circulation performance of 64 Thomson newspapers and 128 non-Thomson papers of comparable size between 1980 and 1990. The Thomson papers lost seven percentage points of household penetration in that period, while those in the comparison group lost only one point.

New management tried to improve the Thomson papers in the 1990s. One innovation was an in-house training program for young people without journalism degrees, or, in a few cases, no college at all. The goal was to produce journalists who could "bring a passion for readers to their work, unencumbered by lofty preconceptions of what journalism is all about," a Thomson executive explained. Also, by training local people who already had a commitment to their communities, the organization hoped to reduce staff turnover.[17]

But by 2000, Thomson gave up on the business and announced plans to sell its newspapers except for the flagship *Globe and Mail* in Toronto.[18] In hindsight, what the company did looks a lot like a harvesting strategy, along the lines suggested by Michael Porter (chapter 1). Lacy and Martin see the case as a sign that demand for newspapers might be inelastic, as Fine suggested (chapter 10), but only so long as some minimum level of quality standard is maintained. When quality sinks below that critical level, demand becomes quite yielding.

"This model of news demand," said Lacy and Martin, "is based on the premise that many readers have a minimum level of acceptable quality that they expect from a newspaper. This minimum is represented by a kink in the newspaper's demand curve. Below the minimum quality

15. Quoted in M. L. Stein, "Publisher Pans Peers," *Editor & Publisher,* February 6, 1988, 9. The other two newspaper groups named by McClatchy were Donrey Media and Lesher Newspapers.

16. Cited in *Editor & Publisher,* May 29, 1993, 14. Two Thomson executives declined to be interviewed for this report.

17. Martha L. Stone, "Thomson Creates Its Own J-School," *Editor & Publisher,* March 20, 1999, 41.

18. "Thomson Corp. to Sell Newspapers," *Editor & Publisher,* February 15, 2000.

level . . . readers will more readily substitute other media as quality continues to decline."[19]

While this is a difficult proposition to prove, it is certainly consistent with the existing body of research on newspaper quality and its consequences. Let us revisit the curve shown in chapter 1 and see how it fits the process suggested by Lacy and Martin. Here is the picture again:

Figure 11-1

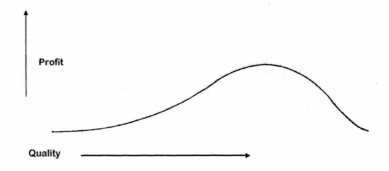

In this model, you will remember, quality has an increasing effect on profitability, up to a point. Then the effect flattens out and turns negative. This theory is nothing more than an application of the law of diminishing returns. The first dollars for quality are spent on the basics such as local coverage, attractive presentation, and a robust news hole. But after a certain point, the additional utility to readers and advertisers is less than the additional cost.

Larry Jinks, former publisher of the *San Jose Mercury News*, finds this model intuitive. "I think that up to a point, what I would define as quality journalism is good business, even in an atmosphere where there is too much attention paid to short-term rather than long-term values. There comes a point where that's probably not necessarily true. You've got 50 reporters in a small town. If you go to 60, is that bound to improve the bottom line? No, not necessarily. Up to a point, quality certainly pays."[20]

19. Stephen Lacy and Hugh J. Martin, "Profits Up, Circulation Down for Thomson Papers in 80s," *Newspaper Research Journal* 19:3 (Summer 1998), 63.

20. Larry Jinks, interview with Jane Cote, San Jose, California, October 25, 2002.

The problem for management and investors alike is defining that "up to a point" spot at the top of the hump. In the absence of a very long-term study, it has to be discovered by trial and error. The managers at Thomson Newspapers evidently thought they were on the downhill portion at the right side of the curve where pulling back on quality increases profit. Instead, they were on the uphill side, where diminishing quality hurt both readership and profit. But by the time they figured that out, the damage was done.

It is an understandable mistake. The effects of quality on readership—and, by extension, profit—are not perceived immediately. Aggressive dilution of the quality of their journalism can make managers look like geniuses to their investors for a while. But, in time, the inevitable price is paid.

The Role of Executive Compensation

Corporations, including newspaper companies, have long believed that they will do better if they can make managers think like owners. The obvious way, giving managers equity in the company, is not as easy as it sounds. In 1994, Congress changed the tax law to prevent executive pay in excess of one million dollars from being deductible to the corporation unless it was tied to performance. That led to a new wave of bonus and stock option plans.

Performance can be measured at both the input level—what specific things the manager does for the company—or the output level, that is, what happens to the company as the result of all factors, only some of which management can control. In my brief time in the executive suite at Knight Ridder, I benefited from both. Knight Ridder was an early adopter of management by objective (MBO), a system that ties some portion of an annual bonus to the performance of specific tasks by the individual manager. I like it for its rationality. Once a year, I sat down with my boss, Jim Batten, to negotiate goals that were clear enough so that a third party could look at the results and know if each goal had been met or not. One of my favorite goals was producing research for the company that was sound enough to be accepted by peer-reviewed academic journals. That met the third-party test, and it contributed to the profession as well as the company.

But the greatest portion of my annual MBO bonus was tied to company, not individual, performance. And company performance is generally measured by financial results, such as operating income, revenue growth, or earnings per share. Results tied to social responsibility factors such as community involvement or product excellence are rare, according to Gilbert Cranberg, Randall Bezanson, and John Soloski, who investigated the effects of public ownership on newspaper companies.[21]

Stock options are the other popular method of tying compensation to performance. An executive is awarded the future right to buy a given number of shares in the company at a stated price, usually the price at the time of the award. It's worth money only if the share price goes up. The problem with this form of compensation is that it can align management interest more with that of short-term traders rather than long-term investors. George Washington University Law Professor Lawrence E. Mitchell has pointed out that short-term and long-term interests would be the same if markets were perfectly efficient. Everyone would have the information needed to judge a company's long-term prospects, and the share price would be based on the discounted present value of that long-term expectation. But the way stock prices swing from day to day indicates that emotion and incomplete information play a large role. "And so," said Mitchell, "even to the economically uninitiated, the idea that stock markets approach any meaningful efficiency seems like a fairy tale."[22]

Because analysts are preoccupied with quarterly changes in earnings per share, managers tend to focus on that number. In a 2002 speech at Northwestern University, Harvey L. Pitt, then chairman of the Securities and Exchange Commission, said corporate officers "should be required to demonstrate sustained, long-term growth and success before they can actually exercise any of their options." Such a requirement, he said, "would abolish perverse incentives to manage earnings, distort accounting, or emphasize short-term stock performance."[23] Companies eventually get caught when they put out bad information, but it can take some time.

21. Gilbert Cranberg, Randall Bezanson, and John Soloski, *Taking Stock: Journalism and the Publicly Traded Newspaper Company* (Ames: Iowa State University Press, 2001), 49.

22. Lawrence E. Mitchell, *Corporate Irresponsibility: America's Newest Export* (New Haven, Conn.: Yale University Press, 2001) 113–14.

23. Floyd Norris, "Pitt's View: Stock Options Can Be Perverse," *New York Times,* April 5, 2002.

In a reward system that pays off for short-term performance, the temptation to take the risk is great. A manager can make the numbers that boost the value of bonuses and stock options and then get out before the bad stuff takes effect. (One of Tony Ridder's contributions was to mitigate such short-term temptations by revising Knight Ridder's management-by-objective system to set multiyear goals. His plan was put into effect in 1996.)[24]

Top executives get the biggest grants, but Knight Ridder democratized the process somewhat when it granted options to buy one hundred shares at the then current price of $54.81 to all seventeen thousand of its full-time workers in December 2000. That's not enough to change incentives very much, but it at least made employees interested in the share price. (It rose above $75 in the next three years, making the options worth $2,000 per employee for those who hung on to them.)[25]

There is a potential moral issue in granting performance bonuses and stock options to news-editorial workers if it motivates them to cheapen the product by providing less service to the community. The American Society of Newspaper Editors, in its 1975 Statement of Principles, said that journalists "should neither accept anything nor pursue any activity that might compromise or seem to compromise their integrity." Editors should, of course, want to keep the company solvent, but they probably don't need stock options to give them that desire.

The various compensation models remind us that managing to get the benefits of quality is difficult even in a steady-state universe. But newspapers are not in a steady state. Their market position is steadily being eroded by substitute technology. The quality issue needs to be examined with that reality in mind.

Climbing Outside the Box

The newspaper managers and the analysts who justify investment in quality only in the presence of competition from other newspapers are overlooking the obvious. The greater justification should be the competition from newer forms of media. But to deal with that requires

24. Geneva Overholser, "Tony Ridder Responds," *Journalism Junction,* Poynter Online, www.poynter.org. Posted February 4, 2003 (retrieved October 31, 2003).

25. "Knight Ridder Gives Employees One-Time Stock Options," PRNewswire, December 21, 2000. Available at *http://www.knightridder.com/happening/otherreleases.html.*

thinking in new ways, climbing outside the box of traditional newspaper management. For a business that has been so successful for so many decades, new thinking is extremely difficult.

One sign of new thinking is the recent move of large newspaper companies into niche publications. Less-than-daily products aimed at ethnic minorities and young people were a major concern of publishers in 2003. Direct mail, free distribution, and Web products aimed at specialized audiences were being rolled out or studied.[26]

The most obvious way to deal with substitute technology is to enter the substitute business. That's harder than it sounds if the capabilities and opportunities provided by the new technology are still being discovered. The Internet can do many wondrous things. Learning how to make its wonders profitable requires a long series of trial-and-error experiments, performed by organizations with a high tolerance for failure. Newspaper companies seldom fit this description. An example of what a more creative culture can do is the stunning move of Amazon.com, announced in October 2003, to put the complete text of 120,000 books—some thirty-three million pages—in a database available for online searching. This "search inside the book" feature was designed, of course, to stimulate book sales by playing on the weaknesses of people like me who are so pleased to find a reference we've been seeking that we forget about the library and order the book on the spot from Amazon. (I fell for it the first time out.)

Newspapers can be innovative. In North Dakota the *Grand Forks Herald* noticed the profitability of eBay and created a local online auction service for its community. The first auction earned ninety-five thousand dollars, a significant sum for a small paper. Other papers started to copy the idea. The goods came from merchants who provided stuff in exchange for discounted advertising. A whole new breed of consultants sprang into being, people who could help newspapers start online auctions. *The News & Observer* of Raleigh, North Carolina, planned to make it an annual event.[27]

While newspapers have been quick to create online versions of them-

---

26. Lucia Moses, "Industry Looks for New Niche in 2004," *Editor & Publisher,* December 15, 2003. Available at http://www.editorandpublisher.com.

27. David Ranii, "Newspapers Stake Claim on eBay Turf," *Charlotte News & Observer,* October 15, 2003.

selves, they can't afford to stop there. They need to think of new applications, things that use the technology to add value. Developing an organizational structure to maximize innovation was a serious issue at the turn of the century. It was not at all clear how online operations could best be integrated with the print version of a newspaper or if they should be integrated at all.

Many of the early efforts were done by subsidiary companies with a firewall between themselves and their print siblings. In some cases, this step was taken, not to stimulate creativity, but as an accounting strategy. A separate company could be spun off and sold to a public eager to invest in Internet startups. Or it could be structured with a tracking stock that would have the same effect with closer control by the parent corporation.

The real benefit had nothing to do with accounting. Protected by the firewall, the online version could be run by people who were free to think anew without being boxed in by newspaper customs and standards. But the cost of such separation was a lack of communication between print and Internet newsrooms and some unhealthy sibling rivalry.

Clayton M. Christensen and Michael E. Raynor, drawing on the work of Clark Gilbert of the Harvard Business School, have proposed a generic solution to the problem of integrating new technology with the old. First, frame the new technology as a threat, and get a frightened top management to commit the dollars needed to created a response. But, once the resources are in hand, get the project out of the building. Shift it to an autonomous organization that will see not a threat, but an opportunity. That opens up the creative flow.[28]

This issue was discussed in 2001 when the Nieman Foundation held a conference at Harvard's Barker Center on the topic "Paying for the Next News." Clark Gilbert was there, and he described his study of the top one hundred newspapers (by circulation) and their Web operations. He found that Web sites that were separate from the newsroom had double the innovation and much higher market penetration.[29]

This result does not mean that print and Web journalists should

28. Clayton M. Christensen and Michael E. Raynor, *The Innovator's Solution: Creating and Sustaining Successful Growth* (Boston: Harvard Business School Press, 2003), 113.
29. Clark Gilbert, "Newspapers and the Internet," *Nieman Reports* 56:2 (Summer 2002), 36.

never interact. Clark's colleague, Rosabeth Moss Kanter, suggested that a newspaper's Web site could integrate its activities with those of the traditional newsroom without being embedded in that newsroom and its culture.

"Integration actually is a very positive model," she said. "Integration doesn't necessarily mean ownership, it doesn't mean control. It means you know how to coordinate and connect activities."[30]

For an example of a successful absorption of disruptive new technology, Kanter described the alliance between Reuters and Yahoo! which worked out as a successful strategy for both. The wire service found a new way to reach customers, and the Web operation gained content. The underlying principle for the old business trying to adapt, she said, is, "You dream your worst nightmare and then invest in it. Figure out what could hurt you and then figure out how to bring that inside."

That step is obvious enough. Her second recommendation is not so obvious. Business people like to plan and script all their moves in advance before rolling out a new product. But in a world that changes so rapidly, it's better to work without a script. "This is improvisational theater rather than traditional theater," she said. ". . . You experiment. These are all small-scale experiments, and you need a variety of them, a lot of them, because you don't know which will work."[31]

The trouble with experimenting is that it costs money. Knight Ridder spent forty million dollars on its ill-fated Viewtron experiment before giving up. When a new generation of managers set up Knight Ridder New Media in 1996, the company again gambled on one large-scale experiment. The prototype for an embedded newspaper online operation was built in San Jose, under the direction of former *Mercury* editor Bob Ingle, and then rolled out to other newspapers in the group. But it wasn't the right model. ". . . I don't think we made mistakes fast enough," Ingle later told Clark Gilbert, who wrote a Harvard case study on the attempt, "and we didn't learn from them often enough."[32]

30. Rosabeth Moss Kanter and others, "Experiences with Internet Journalism," *Nieman Reports* 56:2 (Summer 2002), 36.

31. Rosabeth Moss Kanter, "News Innovation and Leadership," *Nieman Reports* 56:2 (Summer 2002), 31.

32. Clark Gilbert, "Mercury Rising: Knight Ridder's Digital Venture" (Cambridge: Harvard Business School, 2003), 8.

Newspaper companies going into the Internet as a defensive move, according to Gilbert and his former professor Joseph L. Bower, tended to make the same mistake. They "simply reproduced the printed newspaper in electronic form."[33] But new technology usually gets its foothold by serving markets that had not been served before. Personal computers eventually disrupted the market for mainframes, but not until they had filled a previously unmet need, providing cheap word processing and spreadsheet functions for nonexpert users. Their initial competition was not mainframes at all but rather nonconsumption.

The pattern repeats itself across a variety of businesses. New technology creates new customers that the established business, focused on defending its market, tends to overlook. Building on these new customers, the new business eventually goes after the established ones.[34]

It's the same with newspapers. Part of the good news about the Internet is that the people that newspapers are reaching through it are mostly new customers. Here's what Clark Gilbert found:

"Four out of 10 newspaper Web site readers read the traditional print product; two out of 10 actually subscribe. The overwhelming growth is Web readership. More importantly, even where it overlaps, people use the online product much differently than they use print.

"They use the online product as a utility, as a way to get quick access to information that's useful to their lives. The overwhelming net use of all these sites, even the most local, small market ones, is that they create net readership."[35]

And, of course, this net readership comes disproportionately from the younger age groups that are the source of the long-term loss of traditional newspaper readership. Those of us who wish to preserve the social responsibility function of the press by improving its quality need to stop nagging long enough to start looking at the integrated product and not just the portion that is manufactured from paper and ink.

33. Clark Gilbert and Joseph L. Bower, "Disruptive Change: When Trying Harder Is Part of the Problem," *Harvard Business Review* (May 2002), 98.

34. Sean Silverthorne, interview with Clark Gilbert, "Read All About It: Newspapers Lose Web War," *HBS Working Knowledge,* January 28, 2002. Available at *http://hbsworkingknowledge.hbs.edu/item.jhtml?id=2738&t=strategy*

35. Clark Gilbert, "Newspapers and the Internet," *Nieman Reports* 56:2 (Summer 2002), 37.

The influence model can help us with this wider perspective. If a newspaper company's main product is influence, it is important to know how its Web presence contributes to that influence. It is equally important to extend the influence created by the newspaper's brand name to the Web product—or even to radio and television in those cases where so-called convergence strategies have put newspaper reporters on the air. And to track the value of that influence to investors, we need to develop a broad measure that can assess the trust associated with the brand, regardless of platform.

Such a measure is probably not going to be based on variation in content decisions made within the limited ranges available to newspaper editors. The research examined in this book is fairly strong evidence that for content to make a difference, newspaper companies would have to break out of the narrow range—defined by the industry norm of 10 to 15 percent of revenue plowed back into news-editorial costs—in which it is accustomed to operating.

Leo Bogart found pretty much the same thing with his Newspaper Readership Project (1977–1983), even though that didn't stop newspaper consultants from finding work doing endless surveys to find "what readers want." Editors tweaked their content, never found the silver bullet, and readership kept falling. Early in the twenty-first century, there was a new wave of interest in readership with the same aim of finding low-cost adjustments to content to improve readership. The Gannett Company announced a program called "Real Life, Real News: Connecting with Readers' Lives" to emphasize local news, particularly events and trends that affect how readers live.[36]

Like other periodic fashions in the news business, the idea has a good history. Curtis MacDougall noted the importance of "proximity" and "consequence" to news values in the first edition of his textbook *Interpretative Reporting* in 1938.[37] Such a return-to-basics strategy couldn't hurt, but it would need a major boost in newsroom resources to represent a serious effort to reverse the long-term drift away from newspapers.

To justify a quantum jump in news-editorial resources requires a

36. Lucia Moses, "Gannett Papers Will Get 'Real,'" *Editor & Pubisher,* October 6, 2003, 4.
37. MacDougall, *Interpretative Reporting,* (New York: MacMillan Company, 1938), 102.

broader strategy than just slowing the readership decline. If the industry were to embrace the influence model and use it as justification for preserving its social responsibility functions in whatever new media combinations emerge from the great technological disruption, then it might make a difference.

If newspapers did adopt such a strategy, how would they know it was working? A credibility measure that is multidimensional like the 1985 effort of McGrath and Gaziano is clearly the way to go. As we saw in chapter 5, accuracy, especially as perceived by news sources (who tend to be community leaders), is a major predictor of credibility, which in turn leads to robustness in newspaper sales.

There are two serious problems with such a measure. One is the two-step flow. The elites who are quoted in newspapers make their judgments based on how they are treated, and their attitudes filter down to the public. A newspaper that gets too self-conscious about the two-step process might pander to elites at the cost of serving the mass audience.

The problem is that credibility ebbs and flows with things going on in the news. An investor who based an evaluation of a news medium on its credibility would need the patience to look at long-term measures that smooth out the ups and downs brought on by turbulence in the news. Nevertheless, the potential value of being the most trusted source of news and information in a community makes that information worth tracking. It will be especially important as new hybrid media start to compete with local newspapers for the role of most trusted. If an Internet-based service can capture the trust of an audience whose geographic boundaries also define a retail market, it can set up a competitive battle between new and old media that will benefit readers and advertisers alike. Such a medium would have the tremendous advantage of low entry cost, not available to potential print competitors. And the influence model could motivate it to provide the social responsibility functions that have been traditional for newspapers. The last newspaper reader would have a socially useful alternative.

We should also face another possibility. If newspapers abandon their service ethic in order to harvest their market positions, and if the new multimedia competition is slow to pick up that ethic, who or what might become new providers of socially responsible content?

Alternative Journalism

Charitable foundations are already moving in to fill gaps left by short-sighted application of the profit motive. Training of newspaper personnel to enable them to serve society with greater skill is largely paid for by federally recognized charitable organizations such as the American Press Institute, the Poynter Institute, and the National Institute for Computer-Assisted Reporting. The John S. and James L. Knight Foundation has made training for journalists a major initiative.

There's more. Charities have begun direct efforts to pay for news. They either want news media to experiment with unconventional approaches to news coverage or to provide the means for investigative work not covered in the newsroom budget.

The most visible of these efforts has come from the Pew Charitable Trusts based in Philadelphia and the Henry J. Kaiser Family Foundation of Menlo Park, California. Pew supported experiments in civic journalism for ten years and then moved on to other media projects including the Pew Center for the States, which generates stories and reference material to assist reporters covering statehouses.

Some newspaper editors have been eager to get this kind of help, and others, worried about their ability to discern the motives behind the foundation money, have been critical. The foundations are "tucking some new vegetables into your media stew," warned Rick Edmonds in a report for the Poynter Institute. The wariness of journalists has valid historical basis. Paper and ink are powerful tools. News people develop aversive reflexes to fend off efforts of the better-paid people in the public relations industry who want to manipulate them in order to get some of that power.

But the temptation to accept foundation support is very great when the social purpose is clear and the publisher won't come up with the money. The phenomenon is not new, nor do I come to this argument with clean hands. My own career as a journalist includes many examples of foundation support at critical times. My education in social science research methods was provided by the Nieman Foundation when it sent me to Harvard for the 1966–1967 academic year. The following summer, I accepted detached duty with the *Detroit Free Press* to cover the July race riot and its aftermath. As the riot ended, I proposed an expensive survey to learn the causes, using the methods I had learned at Harvard. It required academic consultants, mainframe computer time, and a corps of interviewers. To get it done, Frank Angelo, the managing editor, had to go outside his budget. Whether he asked the Knights for the money (the company was still private then), I do not know. But Angelo found it from outside sources, using his local connections to score a grant from a nonprofit organization, the Detroit Urban League.[38]

The following year, Martin Luther King was assassinated. At that very moment, I was analyzing the results of a survey on attitudes and grievances among black citizens of Miami. This baseline data gave us the opportunity to go back into the field, talk to the same people, and assess the effect of King's death on the civil rights movement. But, of course, we had not anticipated the need for a second survey, and there was no money in the newsroom budget to do that.

One of my Washington beats at that time was the Office of Economic Opportunity. I mentioned the problem to a source there who talked with his boss, Robert A. Levine, the assistant director for planning and research. The outcome was a federal grant to the *Miami Herald* for the field costs of the second survey. It was as flagrant a conflict of interest as you could ever hope to find, but nobody cared. Everybody wanted that survey. After writing the story for the *Herald* and the wire, I did a formal report, in the sociological jargon I'd learned at Harvard, for OEO. Then, on a whim, I sent it to a scholarly journal. It became the lead article in *Public Opinion Quarterly* the following year.[39] Nobody's reputation, neither mine nor the *Herald*'s, was permanently damaged.

National Public Radio is a model often cited as evidence that nonprofit journalism works. While subscriber support is an important component, more than 40 percent comes from foundation and corporate sponsors, according to Edmonds. NPR keeps a policy manual that spells out the limits of permissible relationships with funders. Grants narrowly restricted to coincide with a donor's economic or advocacy interest are not allowed. However, funders can suggest broad themes, and

---

38. Much later, Lee Hills told me that his sources on the Pulitzer committee said the survey was a factor in awarding the 1968 prize for local general reporting to the staff of the *Free Press*.

39. Philip Meyer, "Aftermath of Martyrdom: Negro Militancy and Martin Luther King," *Public Opinion Quarterly* 33:2 (Summer 1969), 160–73.

the news director puts out a wish list of projects that NPR would like to do if funders find them worthy.[40]

That separation between funders and journalists works at least as well as the wall between the news and advertising departments at the newspapers most Americans read. In my study of newspaper practices in the 1980s, editors representing 79 percent of daily newspaper circulation in the United States reported sometimes getting pressure from advertisers that was serious enough to require a newsroom conversation to resolve the issue. One in four said it happened at least once a month. A substantial minority admitted that their readers were sometimes exposed to advertiser influence that was formalized with news copy slugged "BOM" or some other equivalent of "business office must," indicating that it should go into the paper unaltered because it was part of an advertising contract.[41]

So let us be blunt. Allowing charitable foundations to pay for the news might be risky, but it is probably no worse than a system in which advertisers pay for it.

Yet another nonprofit model exists in the form of organizations formed for the purpose of doing investigative reporting. The availability of the Web greatly facilitates this kind of work. Journalists who would refuse, on moral and professional grounds, to rely solely on press releases are often comfortable taking information off the Web without independent investigation, perhaps because it makes them work a little harder than when the handouts are delivered right to their desks.

In 1990, Charles Lewis dropped out of broadcast journalism and opened the Center for Public Integrity in Washington, D.C., as an independent source of investigative reporting. Its work is produced as books, monographs, and online reports. Some of it has won reporting awards from organizations controlled by more traditional journalists, including the 2000 IRE award for outstanding investigative reporting in the online category. The funders included such well-known names as the Ford Foundation plus some foundations that were built on old newspaper money including Times Mirror, Park, and Knight.

40. Rick Edmonds, "NPR: A Firewall, Almost," in *Behind the Scenes: How Foundations Have Quietly Seized a Role in Journalism, Commissioning Content,* Poynter Report, special issue, Spring 2001.

41. Philip Meyer, *Ethical Journalism* (White Plains, N.Y.: Longman, 1987), 40, 205, 236.

Lewis's scope was national. There are others with a more local orientation. Sometimes, they hire former newspaper reporters.

There is, in short, more than one way to pay for the next news. The fact that newspapers are not doing so well these days should not blind us to the possibility that the influence model could reappear in unexpected form. Remember its elements. Advertising gains value, not through interference with the news product, which undermines the long-range interest of everyone, including advertisers, but by appearing in a medium with a reputation for integrity.

How would we create such a medium using the tools and economics of new technology?

First, it would have to be community based. The influence model needs a public sphere defined by both economic and social ties. When my home state of Kansas was surveyed, and the horse was the chief mode of transportation, counties were designed to be just large enough to enable every resident to travel to the county seat, conduct business, and return home on the same day. Today's technology greatly enlarges the potential public sphere, but it does not free it entirely from the constraints of geography. Too much social and economic interaction requires face-to-face contact. Most of us are social creatures who would still rather go to the ballpark once in a while than watch every game on television.

The chief threat to newspapers in the twenty-first century will come from entrepreneurs who figure out how to use the more favorable cost structure of Internet-based media to provide better services to the same kinds of communities that newspapers have served so well. Some of these entrepreneurs might be newspaper companies, but don't bet your career on that. If newspapers harvest their goodwill to maintain their historic profitability, they will create opportunities for entrepreneurs who are willing to try new things and be satisfied with smaller returns.

These rough beasts waiting to be born could surprise all of us. We who wish to preserve the social responsibility functions of the press might do well to turn our attention away from the owners and investors and, instead, look to the people on the front lines who do the daily work of the profession. Whatever form the new journalism takes, it will need a plentiful supply of moral and capable journalists. That need will never change. The next and final chapter explores ways to meet it.

12

What We Can Do

M A Y B E the bean counters will get religion.
Maybe the suits who run media corporations will
give more attention to social responsibility. Let's
not sit around and wait. The time has come to
think about some things that we on the ground
can do while traditional news media struggle
for survival.

What we need is a working-level dedication
to the traditional standards and public service functions of journalism. They're not complicated.

"Get the truth and print it." Two decades
after the death of John S. Knight, his
motto was still cited by his journalists
and their successors, and it continues to
summarize the essential mission of traditional reporting as accurately and
succinctly as anyone could.[1] If he
were alive today, he might rephrase it
with a bow to the Internet: "Get the
truth and post it."

But Jack Knight had another
motto just as important. It was
cited in a memo written by Executive Editor Lee Hills to social-
ize new hires to the company

---

1. Quoted in *Knight Ridder News*,
Winter 1999, 2.

228

culture and handed to me on my first day at *The Miami Herald,* July 1, 1958. "John S. Knight has said," Hills wrote, "there is no excuse for a newspaper to be dull."[2]

In these two aphorisms are both the glory and the pain of journalism today. Of course we have to get the truth and print it, but that's no longer enough. We also have to process that truth in ways that make readers want to go to the effort of receiving it. If that was true in the 1950s, it is even more vital in the rich information environment of the new century.

The values of journalism are under pressure from many sources, not just from the demands of short-sighted investors. If the old journalism fails to adapt, people who know how to use the new technology better than we traditionalists do—or who are just more willing to experiment— will start to crowd us out. At the very least, journalism as a discrete concept with its own set of skills and values is in danger of losing its identity.

From Production to Processing

What was once a steady-state information business is undergoing a paradigm shift comparable to the food business during the development of modern agriculture. As technology has increased the efficiency of farming, the main economic activity has shifted from food production to food processing. In 1947, when the nation's economy was recovering from the effort of World War II, farms contributed $20.2 billion to the gross domestic product while the manufacturing of food and kindred products accounted for less than half that amount, $9.3 billion.

But since 1983, processing has consistently outstripped production in economic value added. In 2001, farming contributed $80.6 billion but food manufacturing was worth $123.7 million.[3] Consumers are no longer concerned with just having enough to eat like the subsistence farmers of the nineteenth century. Now they care about food's taste, texture and nutrients, even the convenience of its packaging.

Technology is pushing information businesses in the same direction.

2. Lee Hills, memo to new employees, 1958. Quoted from memory.
3. National Accounts Data, Bureau of Economic Analysis, U.S. Department of Commerce, www.bea.doc.gov.

Fig. 12-1: Farming v. Processing (Constant 2001 Dollars)

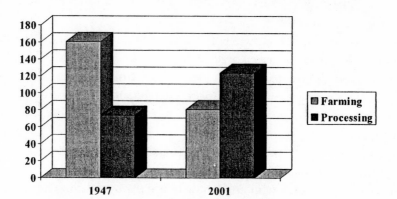

With information so plentiful, skill at finding and delivering the truth becomes relatively less valuable while emphasis shifts to the ability to make the product attractive and desirable to the end user. Editing is growing in importance relative to reporting. The best-paid graduates of journalism schools today are the Webslingers, young people who can design and package visual displays for information and construct the hyperlinks that hold different elements together. In the journalism graduating class of 2001, those working in online publishing earned an average of $33,500 a year compared to $25,896 for those who started at newspapers.[4] (The gap has since narrowed as the supply of online editors has increased.) The importance of design and packaging is not confined to the new media. The creation of *USA TODAY,* an editor-driven paper, was a successful early adaptation of traditional print to these new needs.

The new emphasis on filtering, refining, decorating, and packaging information requires new ways of applying our old skills and discovery of some entirely new skills. It is no coincidence that Richard Curtis, designer of *USA TODAY's* way of presenting information, is not a journalism school product but a graduate of the School of Design at North

4. Lee B. Becker, Tudor Vlad, Jisu Hu, and Nancy R. Mace, "2002 Annual Survey of Mass Communication Graduates," James M. Cox Center for International Mass Communication Training and Research, University of Georgia, 2003. Online salaries declined in 2002.

Carolina State University. As the demands upon us are new, we must think anew.

Demand for new thinking is moving control down to younger age levels. The need for some of the new skills, such as Web page creation and editing, is so intense that the jobs are being filled by nonjournalists or by new journalism school graduates who have not obtained the socialization process provided by traditional media to transmit the craft's values. Meanwhile, in the cluttered information marketplace, where information itself is no longer scarce and therefore less valued, the attention of the public has become the scarce good.[5]

In order to gain that scarce attention, professional communicators are trying a number of things that, while not performed by real journalists, are not always distinguishable from journalism in the public mind. One is to make content as outlandish and shocking as possible. Another is to get information into print, on the air, or on the Web so quickly that there is no time for fact checking. Yet a third is to blend editorial content with paid advertising or public relations material so seamlessly that the consumer is unaware that he or she is receiving a commercial pitch. Even though these things are not normally done by legitimate journalists, there are enough highly visible infractions by real journalists to make the distinction fairly seamless in the public eye.

My thesis in this chapter is that journalism's response to these new demands will slowly but irreversibly force it to move from craft to profession. It is time to band together for self-protection and for clearer identification. But does what journalists do really meet the definition of a profession? Not in every respect. But we are starting to get very close.

Professional Education

One of the defining characteristics of a profession is access to a highly specific, even arcane, body of knowledge.[6] Not everyone in the traditional media is convinced that journalism has a body of knowledge. This conclusion would surprise the publishers of textbooks on reporting,

5. Originally articulated by Herbert A. Simon in 1970, this point has been widely applied, for example by Neuman in *The Future of the Mass Audience*.

6. Wilbert Moore, *The Professions: Roles and Rules* (New York: Russell Sage Foundation, 1970), 6.

editing, media law, ethics, computer-assisted reporting, and other relevant topics. But those who deny the body of knowledge tend to be the same people who deny the need for journalism education. Many newspaper editors still claim to prefer to hire liberal arts graduates instead of those with degrees from schools of journalism. Their practice shows otherwise. A 2000 survey found that 73 percent of new hires at newspapers were journalism school graduates. Media companies lack the patience to train totally fresh recruits, and they have learned through experience that the J-schools are a reliable source of cheap labor. The convergence of media will increase their dependence on journalism education because of the greater need for specific technical skills. In TV news, where knowing how to work the equipment is a prerequisite, 94 percent of new hires came from J-schools.[7]

Once the journalist's mission shifts its emphasis from transportation to processing, the body of knowledge becomes more intricate and specialized. If the mission is to get the truth into the heads of the audience, not just in its hands, then an understanding of the cognitive processes that news consumers use becomes important. Suddenly all of those so-called conceptual courses in journalism schools on the processes and effects of mass communication start to assume new relevance.

Professional status also suggests a moral concern for the effects of mass media that is not assumed by the craft model. One way to understand the war of words over civic journalism is as a battle between the new professionals and the old craft persons. The old ones tend to be nonconsequentialists. They see their job as to get the truth and print it, period. Let the chips fall where they may. The craftsman needs neither to know nor care where the chips fall.

The civic or public journalist, as Buzz Merritt has explained so well, does care. An appreciation for the effects of his or her work is part of the definition of a public journalist. Caring, as Merritt is quick to point out, doesn't mean being manipulative. But it does mean choosing topics and modes of reporting and writing that lubricate the sometimes

7. Lee B. Becker, Tudor Vlad, Robert A. Papper, and Michael Gerhard, "Survey of Editors and News Directors," August 6, 2001. Retrieved in January 2003 from www.grady.uga.edu/annualsurveys.

abrasive process of public judgment. A mere transporter of information can't do that.[8]

Traditional journalists are quite skeptical about this self-conscious agenda setting. Jonathan Yardley argued in *The Washington Post* that the concept distorts the journalist's decision-making process "and offers him insidious possibilities for twisting the public debate to his own liking."[9] Such opposition is, of course, an ironic affirmation of journalism's status as an incipient profession. One of the less admirable things that professionals do is exercise their social control to resist the development of new and threatening technologies and value systems.

Wilbert Moore was talking about medicine, not journalism, when he observed thirty years ago how professionalization becomes a conservative force against innovation. His words still make a good fit to the emotional opposition of establishment journalists to Merritt's proposals. "What is of enduring importance," Moore said, "is the homely truth that new knowledge or innovations in technique and practice threaten the very basis upon which established professionals rest their claims to expert competence."[10]

A profession has clients. Who is the journalist's client? It is not the news source, although many sources think so. It is not the advertiser nor the publisher and shareholders. Traditional journalists consider the reader as their client. The desire of the civic journalists to emphasize the collective nature of the client makes perfect sense. Readers' interests cannot be taken into account at the level of every single individual, and so journalism's client is the community. A news medium can treat its community of users with the same care and concern that a teacher shows for a student or a physician for a patient. Civic journalism encourages media to think of society as the client, and, to the extent that this idea catches on, the service component of professionalism takes on new meaning.

There has always been such a component, of course. From the time of John Milton, the power of truth put to public service has been a main

8. W. Davis "Buzz" Merritt, *Public Journalism and Public Life: Why Telling the News Is Not Enough*, 2nd ed. (Mahwah, N.J.: Lawrence Erlbaum Associates, 1998).

9. Jonathan Yardley, "'Public Journalism': Bad News," *Washington Post*, September 30, 1996.

10. Moore, *The Professions*, 44.

motivating factor for journalists. Jack Knight's first motto still draws idealistic young people to the profession despite its low pay.

Setting and Enforcing Standards

But to be a true profession, an occupational class has to be organized to uphold and advance its standards of service and truth telling. The motivation need not be entirely altruistic, because the economic basis of a professional relationship is trust—both in the professional's moral values and in his or her technical competence. So a key function of a professional organization is to set both moral and technical standards. Attempts to enforce moral standards by professional bodies are seen only sporadically in journalism. Setting technical standards has been attempted by employers through the testing of prospective employees, but never by a professional body. That could change soon. The new demands of certain subsets of skills, such as computer-assisted reporting, have led to some discussion of competence standards.

On the moral side, associations of journalists since the 1920s have experimented with codes of ethics as one way to gain and keep that trust. But efforts to make codes enforceable have been mostly ineffective. The American Society of Newspaper Editors tried to enforce its 1923 code in 1926 by censuring a corrupt member but backed down under the threat of litigation.[11]

Those who resist having enforceable codes often cite advice that lawyers give to media companies. Since the Supreme Court extended protection from libel suits in *Times v. Sullivan,* litigants who are public figures have been forced to prove malice on the part of the journalist in order to win their cases. That was good news for journalism, but it came with some accompanying bad news: reporters could be forced to give up their notes or made to testify about newsroom conversations so that litigants could search the contents for evidence of malice. And action taken in contravention of a news organization's own code of ethics could be interpreted as malice. Better, say the lawyers, to have no explicit moral standards than to have your failure to apply them used against you in court.

11. The case is recounted by Bill Hosokawa in his history of the *Denver Post, Thunder in the Rockies* (New York: Morrow, 1976), 146–47.

There are other arguments. One is that enforcing codes of ethics would interfere with the First Amendment rights of journalists. The other is that ethics in journalism is such a slippery subject that enforcement would be unworkable. For every rule, one can always envision a situation where it ought to be broken. Therefore, goes the argument, there can be no rules. Yet, there is some experience that tells us otherwise.

One of the most important precedents is more than a century old. The Standing Committee of Correspondents of the House and Senate press galleries was organized in the nineteenth century (1879 for the House, 1884 on the Senate side) precisely for the purpose of distinguishing between real journalists and advocates for narrow interests. Over time, it gradually assumed, albeit slowly and reluctantly, the role of judging ethical behavior as well.[12]

The power of the Standing Committee comes from its responsibility for issuing credentials admitting reporters to the galleries. These credentials may not be as important as they were before Brian Lamb started C-Span and enabled reporters to cover Congress from their TV sets, but its privileges can be critical for a reporter's career advancement. The staff of the Press Gallery issues press releases, provides work space and telephones, maintains a library of C-Span videotapes, and relays messages for reporters. A member of the Press Gallery can invite a senator participating in floor debate to the President's Lobby or a House member to the Speaker's Lobby for a personal interview. The Press Gallery card is still the basic credential for Washington journalists and is a prerequisite for a White House press pass.

The Standing Committee is not a professional association in the sense that it is empowered solely by its membership. The power to determine who sits in the galleries remains with the two houses of Congress. But it was recognized long ago that the decision should be delegated to a system of peer review.

And so members of the Standing Committee are elected by their peers. The rules have always been simple, although not always fair. Prior to World War II, the rules were employed in a discriminatory way to keep blacks and women out of the press gallery, although one black journalist, Frederick Douglass, was admitted as editor of the *New National Era*

---

12. Donald A. Ritchie, *Press Gallery* (Cambridge, Mass.: Harvard University Press, 1991).

in the 1870s. The next black person was admitted in 1947. The rules
have always, however, been simple. Admission is limited to "bona fide
correspondents of repute in their profession," which are defined for the
daily press as follows:

a. Their main income is obtained from news correspondence intended for newspapers entitled to second-class mailing privileges.

b. They are not engaged in paid publicity or promotion work nor in
prosecuting any claim before Congress or another department of government.

c. They are not engaged in any lobbying activity.

There are parallel rules for reporters in broadcasting and periodicals,
and for photographers. What they all have in common is an attempt to
define with some precision the nature of a "bona fide" correspondent.
For the periodical galleries, for example, "applicants must be employed
by periodicals that regularly publish a substantial volume of news material of either general, economic, industrial, technical, cultural, or trade
character. The periodical must require such Washington coverage on a
continuing basis and must be owned and operated independently of
any government, industry, institution, association or lobbying organization."[13]

Wilbert Moore's observation about professional societies resisting the
new technologies holds true in the history of the Standing Committee.
Magazine writers, then radio and television reporters, and most recently
reporters for Internet services were admitted only reluctantly. But the
rules are more than aspirational codes. There is an enforcement procedure, and it is employed, although not often.

In 1979, the Washington bureau chief of *The Detroit News*, Gary F.
Schuster, was reprimanded by the Standing Committee for impersonating a government official to get a story. Schuster's stunt was to get access
to a White House ceremony by impersonating Michigan Representative
Bob Traxler and riding the special bus for members of Congress. He
could have used his White House press pass to get into the ceremony,
but he wanted to make a point about lax security in Washington.

13. http://www.senate.gov/galleries/pdcl/rules.htm.

While the Standing Committee lacks an explicit code of ethics, chairman Michael Posner of Reuters news agency said that the responsibility for issuing press cards includes the "responsibility to see that credentials are not abused."[14] A similar incident took place in 1989 when columnist Jack Anderson interviewed Senator Bob Dole inside the Capitol and displayed a gun and a bullet that he had smuggled into the building to demonstrate lax security. He, too, was reprimanded.

These cases illustrate an important function of a professional organization. When nonprofessional behavior comes to public attention, somebody needs to speak for the profession and say, in effect, "This is outside the bounds of our normal and approved behavior; this person's actions do not define us." Thus it sharpens the boundaries between legitimate journalists and all the others. This sharpening of boundaries helps journalists with their own struggles for self-identification, but, more importantly, it helps the public with that same definitional problem.

This function does not depend on an enforcement mechanism that identifies and punishes every professional misdemeanor. All it requires is a few visible cases. An aspirational code comes to life every time it is held up against a real case and a judgment rendered by a body of peers elected to represent the profession.

The Minnesota News Council performs just such a function, although it includes mixed representation from both the profession and the community. But it, too, does its job just by making and publicizing a judgment. Could broader organizations of professional journalists do the same?

The Pollster's Case

A professional group with close ties to journalism sets a useful example. The American Association for Public Opinion Research represents both academic, commercial, and nonprofit pollsters. It was founded in 1947, the same year that the Hutchins Commission published its report, and its members are required to sign its code of ethics.

14. Quoted in Lois A. Boynton, "The Dark Side of Press Gallery Accreditation," Southeast Colloquium, Association for Education in Journalism and Mass Communication, Lexington, Kentucky, March 1999, 18.

Public opinion polling started out as an arm of journalism. The first straw polls were done by newspapers to try to predict election outcomes. George Gallup made his fame and fortune by publicly challenging the unscientific poll of *Literary Digest,* a national magazine, with a better one that he freelanced to newspapers. That was in 1936. He later became the eighth president of AAPOR in 1954–1955.

Pollsters have the same basic mission as journalists: to discover and impart the truth. Like journalists, they use methods that are a mixture of science and art, although they lean more toward science. One consequence of scientific method is that its results can be evaluated by more than mere intuition. It also carries a tradition of openness to promote that evaluation. Its discovery methods are more structured than those of standard journalism, and this structure makes it easier to develop rules of conduct and to know when they are violated.

AAPOR's code developed slowly. A committee on standards was formed at its very beginning, but it did not adopt a code of ethics until 1960. The procedure for dealing with violations was set up in 1975. It was applied in that same year with a finding of bias in a poll that Opinion Research Corporation had executed for an interest group that opposed the establishment of a consumer protection agency in the federal government.

In 1991, AAPOR censured Planned Parenthood of Minnesota for using a poll to assemble a database of names and addresses of potential supporters without telling the respondents that their names would be used for nonresearch purposes. In that same year, a similar finding was made in a case involving the National Right to Life Committee.

More recently, AAPOR censured Virginia pollster Frank Luntz, who published results of a poll taken on behalf of Republican political candidates but declined to disclose the methodological details. Luntz was not a member of AAPOR, so the association could not expel him, but that is not its practice. In keeping with the libertarian theory of the press, it sheds light on wrongdoing and expects the publicity to have a healing effect. Publicity has always been its only sanction.

Why can't journalists do the same? The First Amendment is often cited as an objection, and it was raised in litigation against the Standing Committee of Correspondents, but not successfully.

Vigdor Schreibman, a retiree who operated an online news service out of his home in the Capitol Hill neighborhood of northeast Washington, was denied a press card in 1996. The Standing Committee decided that his operation was not a genuine news medium. He lived on his retirement income, his publication lost money, and its volume of material was deemed not substantial. In his appeal, Schreibman argued that most Internet publications lose money, and he claimed that his First Amendment rights were violated because he could not compete effectively with other media without the services provided by the Press Gallery. A federal appeals court disagreed, and, in February 2000, the Supreme Court refused to review.

AAPOR's case is even less of a free-speech issue. While its public resolutions of censure might have a chilling effect on the specified practices, they do not prevent their subjects from continuing them. All AAPOR does is expose them to public view—which is exactly what First Amendment theory says should happen. John Milton, who in 1644 argued for letting truth and falsehood grapple in a free and fair encounter, would have approved.

Even then, exposing or discouraging bad practices is not the most important effect. The main function of the censure resolutions is to draw a public distinction between legitimate polling practitioners and those outside the pale. And this is exactly what journalism in general needs if it is to retain its professional identity. Drafting codes is easy for journalists. So is arguing about their application in specific cases—the ethics listserv of the Society of Professional Journalists buzzes with debate over visible ethical controversies. But SPJ has no formal enforcement process like AAPOR's, and the debated issues are never resolved. For professional journalism to build its identity, it needs to come to grips with at least a few specific cases and use them to test the meaning of the aspirational codes.

There have been some recent steps in this direction. In 2000, the board of the Minnesota professional chapter of the Society of Professional Journalists publicly criticized a television reporter who removed a tape of an illegal dog fight from the owner's car, copied it, and turned the original over to authorities. "Professional journalists cannot and will not condone these types of actions in pursuit of this story," the board said, citing four broad provisions of the SPJ code that were violated.[15]

15. David Chanen, "Journalists' Group Criticizes Reporter's Actions," *Star Tribune,* May 13, 2000.

In 2003, the national SPJ office issued a press release critical of "some broadcast stations" for the practice of getting business people to pay to be interviewed on the air. "Some of the recent offenses have taken place in broadcast programs that fall somewhere between news and entertainment. It is the duty of the ethical broadcaster and print journalist to assure that there can be no confusion in the public's mind which content is advertising and which is news or opinion or entertainment," the Society said.[16] The statement did not identify the perpetrators or their stations.

The American Society of Newspaper Editors has also taken action against a publisher in a very special and limited case. The publisher was the acting president of Hampton University who seized copies of a student newspaper. ASNE, in a powerful pocketbook sanction against the university, withdrew its financial support for a program training high school journalism teachers.[17] If the frequency of such cases increases, these or other journalism organizations might eventually feel the need to formalize the sanctioning process, and another step toward professionalization will have been made.

Certifying Competence

The other side of professional definition depends less directly on moral issues and more on competence. Of course, competence itself can be a moral issue if a practitioner pretends to competence that he or she does not possess. In simpler times, when the reporter was a hunter-gatherer of information, the presence or absence of competence was highly visible. One either brought home the goods or returned empty-handed. Unlike those of a surgeon or attorney, the mistakes of a journalist were out there in public for everyone to see and notice.

But that transparency diminished as journalism began to require more technical competence. The aphorism "A good reporter is good

16. "SPJ Calls on News Media to Maintain Clear Separation of News and Advertising," press release, November 10, 2003, http://www.spj.org, retrieved November 12, 2003.

17. The ASNE announcement took the form of a November 11, 2003, revision to its October 7 press release announcing the high school program. http://www.asne.org, retrieved November 12, 2003.

anywhere" no longer holds true when reporting uses specialized knowledge. One growing subset of journalism that requires technical skill illustrates the problem.

In the late 1960s, computer software developments began to make computing tools more accessible to nonspecialists. Arthur Couch, a sociologist at Harvard University, wrote Data-Text, a higher-level language based on FORTRAN and the assembler code of the IBM 7090 to make it easy for a nonprogrammer to analyze data with crosstabulations, correlations, and their related statistical tests. It was only a matter of time before journalists would find out about these tools and apply them in their own work.

By 1989, enough journalists were making use of such applications to form a professional organization. The National Institute for Computer-Assisted Reporting (NICAR) was created as a program of Investigative Reporters and Editors (IRE), which had been formed in 1975. Its primary focus was on training and sharing of ideas. But as computers became cheaper and more powerful and as the number of applications expanded, the problem of occupational definition emerged yet again.

In 1999, the Poynter Institute held a tenth anniversary meeting of some of the key players in NICAR and the CAR movement in general. By then, the definition of CAR had expanded to include a great variety of computer applications, including online searching. There were so many different people doing so many different things under the CAR label that its identity was no longer clear. However, three major categories could be identified. Ordered by level of difficulty, they were:

1. Online searching and data retrieval
2. Sorting and classifying data with spreadsheet and database tools
3. Statistical analysis and application of scientific method

Already it could be seen that an editor seeking to hire a CAR reporter needed some way of discovering which of these skills a self-identified CAR specialist possessed and how good he or she was at them. The topic of certification came up.

The first reaction of the veteran journalists assembled there was one of dismay. Certification seemed to be at odds with our libertarian tradition. It sounds like a step toward licensing. But, on further examination and discussion, some possible benefits were identified and acknowledged.

A certificate is basically a piece of paper that says some recognized agency has examined a person's ability and found him or her qualified in certain areas and at certain levels of skill. A high school diploma is a certificate, and it is one whose utility most employers, including the U.S. Army, recognize. If nothing else, it sorts the good risks from the bad. At the same time, it does not make it impossible for the nonholder of a diploma to get a job, and it certainly doesn't make it illegal. It is a way of communicating information that follows a standard definition.

If we think of certification as a form of communication, then it makes quite a good fit to our libertarian leanings and our desire for openness. Communication requires language, and to use language we need definitions. A certification program would enable a job applicant to demonstrate to a potential employer some concrete and instantly understandable evidence of a specified level of skill. Computer professionals have already found this concept useful, and a number of private training institutions have sprung up to create certification programs in specific computer skills.

Journalism schools are already under pressure to provide midcareer training so that those who graduated before the computer's use became so common can feel less disadvantaged in comparison to new, computer-ready graduates. A certification program would be a logical part of a midcareer training program. And both the schools and the midcareer students should be comfortable with it since a journalism degree is itself a form of certification. So, for that matter, is a passing grade in any specific skills course.

Cultural Change

In these two areas, certifying competence and enforcing rules of ethics, the nascent profession of journalism stands ready to emerge. The old culture of journalism that has resisted such change is already starting to yield. After thirty years of operation, the Minnesota News Council is no longer considered a deviant case. It has been endorsed by such old-school journalists as Mike Wallace, Gene Roberts, Bill Moyers, Geneva

Overholser, and Hodding Carter.[18] The Gannett Company rewrote its internal ethical rules after a highly visible and embarrassing retraction of *The Cincinnati Enquirer's* investigative story about the Chiquita Banana Company. A logical next step would be for media companies with codes to enforce them in a public and visible way. But an even better step would be to have professional societies not tied to any single company doing the job.

Who will step up and volunteer? Specialists in fields that are easy to define but hard to learn would make good candidates. In 1998, the medical editor of ABC News, Timothy Johnson, made a compelling argument for certification of medical journalists.

"Unlike the reporting of standard news, which requires general journalistic skills and familiarity with the subject matter," he said, "good medical-news reporting requires additional and very specific skills in the understanding of biostatistics and epidemiology."

Johnson, who is a physician and holds a master's degree in public health, said not all medical journalists would need as much formal training as he has had. But he argued for "some kind of system to ensure that those who wish to become medical journalists have a basic knowledge of the subject and some way of certifying them that would be recognized by employers and the reading and viewing and listening public."[19]

A precedent exists in television. Many TV weathermen are certified as meteorologists by the American Meteorological Society. Getting accurate information about developments in medicine is surely at least as important as getting reliable weather information.

Biologists and social scientists alike are starting to agree that moral systems are formed and persist because they have survival value for the social groupings that create them.[20] Journalism's traditional value set was based on the economic and mechanical constraints of the newspaper

18. All but Mike Wallace made their endorsements in an ad paid for by the Ethics and Excellence in Journalism Foundation, *American Journalism Review,* January/February 2000. Wallace announced his change of heart much earlier.

19. Timothy Johnson, Shattuck Lecture to the Annual Meeting of the Massachusetts Medical Society, May 9, 1998. *New England Journal of Medicine* 339:2, 87–92.

20. For example, Francis Fukuyama, *Trust: Social Virtues and the Creation of Prosperity* (New York: Free Press, 1995); James Q. Wilson, *The Moral Sense* (New York: Free Press, 1993); Edward O. Wilson, *Consilience: The Unity of Knowledge* (New York: Knopf, 1998);

business. New information technology is forcing us to experiment with new ways of working, and that necessarily means experimenting with new ways of defining and organizing our occupational specialties. Professionalism is a higher form of organization toward which the increasing and more complex responsibilities of journalism will inevitably push us. It is a necessary condition for our survival, but by no means a sufficient one. Having well-qualified workers does no good if industry won't pay enough to attract them. In 2002, entry-level newspaper salaries declined in current as well as inflation-adjusted dollars.

The corruption of professional functions by corporations and partnerships has become quite visible in the more established professions such as accounting and medicine. A professional is a flesh-and-blood person who can empathize with his or her customers and suppliers and feels the need for social support in the community. A corporation possesses the legal characteristics of a person but has "no soul to be damned and no body to be kicked" and therefore lacks the humanistic concerns of a real person.[21] But if business—including the news business—is going to be reformed, the initiative should come from those souls and bodies who toil in the field with professional responsibilities in mind.

If journalism is to survive, it will need a professional apparatus as one of the tools in the fight. Trying to reform investors, editors, and publishers is a good idea, but let's not wait for those people to change their ways. Those of us who practice or teach journalism at ground level will make progress with greater speed and certainty if we also organize to reform ourselves. If we can do that, then the next generation of journalists will be ready to work when the process of natural selection chooses the new media forms where trust and social responsibility prevail.

21. Lawerence E. Mitchell, quoting "an English jurist" in *Corporate Irresponsibility: America's Newest Export* (New Haven, Conn.: Yale University Press, 2001), 43.

Afterword

W H E R E do we go from here? The problem of preserving quality in journalism gets especially close attention during cyclical economic downturns. But the underlying problem is not cyclical at all. The business cycle can attenuate or exaggerate the trend, but the long drift toward more specialized media at the expense of mass media seems likely to continue. This trend affects more than just our traditional media. It involves our ability to maintain a unified political culture with shared values. The Internet did not create this problem, but it is accelerating it.

Newspapers have done as well as they have in recent years by finding ways to meet more specialized needs inside the framework of the umbrella newspaper. Examples include geographic zoning, special sections, foreign language editions, and less-than-daily products directed at niche audiences.

The influence model is applicable to each of these potentially profitable niches. Newspapers will always do better in places where they are trusted. The major strategic issue is to discover and understand the specialized populations where it is most feasible to build trust and exercise influence.

Newspaper publishers might believe that the abnormally high profit margins that they enjoyed relative to other businesses in the twentieth century are their birthright, but they're not. They were the result of a condition that no longer exists: their near-monopoly control over retailers' access to their customers. It was a natural monopoly because of the high cost of a printing press. That monopoly has been disrupted by technology that creates cheaper means of distributing information. High-quality journalism will still be economically feasible, but it won't be as profitable. The problem is not one of maintaining the old profitability. That can't be done in a sustainable way. The real problem is adjusting to profit levels that are normal for competitive markets.

Cutting quality to maintain the accustomed profitability can postpone that adjustment, but such a strategy is very dangerous for existing companies. Lower quality will erode trust in the newspaper and create opportunities for what the business schools call bad competitors. What is a bad competitor? One who is willing to provide better service to customers at lower margins of profit. From the viewpoint of society, of course, the arrival of the "bad" competitors might be a very good thing indeed.

To succeed, they will need to find a way to capture, package, and sell the trust that the old media are voluntarily abandoning through their harvesting strategy. The media companies that understand the influence model and apply it to the new, more specialized marketplaces could start to look very much like the ones run by the philosopher-kings of twentieth-century newspaper publishing. And they would provide creative outlets for the best and the brightest of the next wave of young journalists. Watch for them.

Appendix
Some Notes on Data Analysis

C O R R E L A T I O N does not prove causation, but it can be an important piece of evidence. Because correlation-based statistics are used throughout this book, a brief explanation might be helpful for readers who are not familiar with them.

The basic requirement is to have two variables that can be measured on some continuum. Not every interesting thing can be measured in that way, but many things important to business people—dollars and cents, for example—do meet the requirement. Here is one example: newspaper advertising rates and circulation.

Figure 1: Advertising rates by circulation

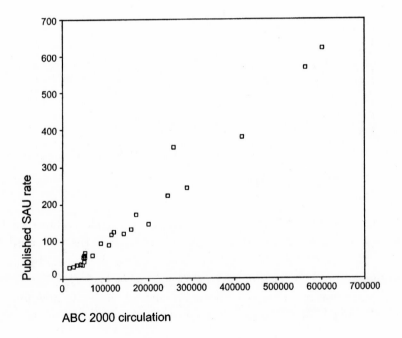

ABC 2000 circulation

247

Are they related? Common sense tells us that they should be, and we can check by drawing a plot that shows the position of different newspapers on the two dimensions. In this example, circulation is measured along the horizontal axis and advertising rate (expressed in Standard Advertising Units) along the vertical. They form a pattern that looks a lot like a straight line.

Using a method developed by Karl Pearson (1857–1936), we can find the line that best describes that pattern. It is called the least squares line because it minimizes the vertical squared difference between each newspaper and the line. Here's the plot again with the line drawn in:

Figure 2: Advertising rates by circulation with regression line

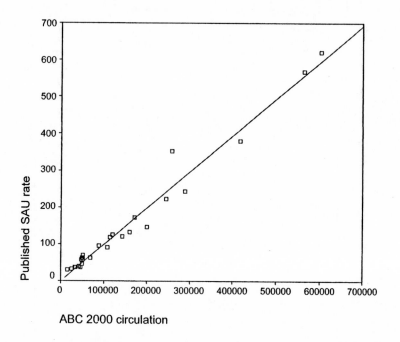

ABC 2000 circulation

This line helps us in a couple of ways. Its slope tells us how much advertising rates increase for each unit of circulation. The value of the slope is .001, meaning that published ad rates tend to go up by one-tenth of a cent for each unit of circulation.

The amount of scatter around the line provides a cue to how much error we would make if we follow the tenth-of-a-cent rule to predict the ad prices of individual newspapers for which we already knew the circulation. If there is no scatter around the line, that is, all the newspapers are spot on, then prediction would be perfect and Pearson's correlation coefficient would be 1. For this particular case of newspaper circulation and ad rates, it is almost 1. To be exact, it is .985.

When we square the correlation coefficient, we have a measure of the variation in SAU rates that is predicted or "explained" by circulation. The square of .985 is .97. In other words, having the information about circulation enables us to guess the ad rates of papers in the sample with 97 percent greater accuracy than we would obtain by using the mean for our estimate.

Predicting ad rates from circulation is no great achievement, but the variation that is left over might be interesting. Pearson's method lets us see it in picture form. Look at the data point that is farthest above the line (at about 258,000 circulation and between $300 and $400 on the SAU scale). According to the regression formula, that paper's managers ought to be asking about $250 per SAU. Instead, they get away with around $100 more. If we could find out how they do that, we might have some useful information for the newspaper industry.

The difference between the expected value (based on the regression line) and the observed value is called the "residual." Think of it as leftover variation after the explanatory power of circulation size has been all used up. We can use that information by taking the residuals for all the papers in the plot and use them as the variable to explore in a new plot with a new explanatory variable. Pearson figured out how to do this with many different explanatory (or "independent") variables in a procedure that was difficult in his day but is easy with a computer. It is called multiple regression, and it estimates the effects of different explanatory variables after the effects of the others have been taken into account.

As you perhaps have noticed at this stage in your life, God did not make the world in straight lines. Consider, for example, the effect of market size on volume of news-editorial content.

The amount of news increases steeply with market size until size exceeds four hundred thousand households. After that, it levels off. If we want to use the linear model, we have two choices. We can treat the markets with less than four hundred thousand as a separate population

Figure 3: News hole by number of households in home county

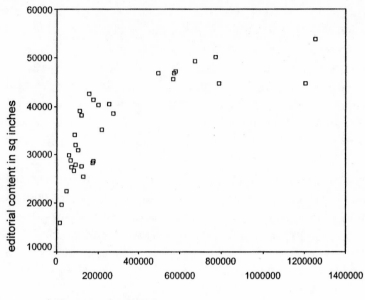

Households in 2003

and set the others aside. Or, we can reexpress market size as its loga-rithm. The advantage of the latter strategy is that we can still use straight-line statistics, but with recognition that what we are really deal-ing with is a curve—and a very specific curve at that.

Here's a look at the same plot with the reexpression:

Figure 4: News hole by log of households

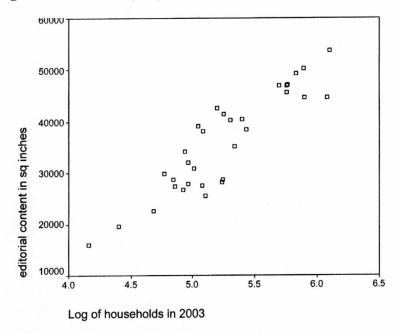

That's much closer to a straight line. Before computers made this so easy, we made plots like this by hand on log graph paper. On the base-10 log scale, the values from 1 to 10 occupy the same linear space as those from 10 to 100 or those from 100 to 1000. In other words, the scale is stretched out on the low end and compressed at the high end. It's as though the previous graph were printed on a rubber sheet, and we stretched the left and compressed the right to straighten the line.

Isn't this cheating? No. It would be cheating to pretend that the original line was straight. Now we have a cleaner mathematical description of the world as it really is. And the correlation coefficient is .907, which is about as good as it gets. The log of households explains 82 percent of the variance in news hole.

When we use regression methods, we are trying to explain variance around the mean of the dependent variable. We do it by looking for covariance—the degree to which two measured things vary together. If we know a lot about these variables and the situations in which we find them, we might even start to make some assumptions about causation.

But correlation is not by itself enough to prove cause and effect. No statistical procedure can do that. In the end, we are left with judgment based on our observations, knowledge, and experience with the real world. We still have stuff to argue about. But with the discipline of statistics, as Robert P. Abelson has said, it is "principled argument."

Because most of the data examined in this book uses continuous rather than categorical measurement, my arguments are based mostly on correlations. Usually, I report the variance explained (r^2) in the text, because it is a more intuitive figure, and put the correlation coefficient (r) in the footnote. Also left to the footnote in most cases is the statistical significance of the correlation, expressed as the probability that the relationship is due to chance.

Sometimes when trying to determine the direction of causation, I use partial correlations. A partial correlation estimates the relationship between two variables after the effect of a third variable that influences them both has been accounted for. This procedure sometimes helps because it can rule out the possibility that some third variable is causing the relationship between the two under investigation. For example, market size is a fertile variable that can make the many things that it causes seem to be related on their own when they are really independent.

When dealing with correlation, it's always a good idea to look at the scatterplot to see if the result is being driven by one or two eccentric cases or if the relationship is nonlinear. This is especially true for small samples like those used in this book. I have included many scatterplots for that reason, as well as to give the reader an intuitive appreciation of the argument that community influence and newspaper success are related.

Acknowledgments

F U N D I N G for the research and time off to do the writing came from the John S. and James L. Knight Foundation along with the specific encouragement of John Bare, Hodding Carter and Eric Newton. And, of course, the memory of the Knight brothers and their ideals guided its development.

Many persons at the School of Journalism and Mass Communication at the University of North Carolina at Chapel Hill helped out along the way. Nancy Pawlow, the Knight Chair assistant, kept the project organized. Many student assistants lent a hand, including Minjeong Kim, Theresa Rupar, Owen Covington, David Freeman, Chris Richter, Joy Buchanan and Cary Frith. Other students let me borrow from their course work, including Koang Hyub Kim, Diana Knott and Yuan Zhang. Fred Thomsen kept my computer running. April Umminger of *USA TODAY* was a summer volunteer. Special thanks to Dean Richard Cole for providing the work environment and bringing in the sheaves.

Colleagues at Chapel Hill and other institutions were a major help, especially Jane Cote of Washington State University at Vancouver whose research interests in her field of accounting overlapped mine in journalism. We collaborated on the interviews of present and former newspaper company CEOs. Scott Maier of the University of Oregon pitched in on data collection and study design for the accuracy measurements. Dale Peskin of New Directions for News managed the contracting on that effort. Frank Fee of Chapel Hill collaborated on the copy editor study. Edward Malthouse at Northwestern University, Rick Edmonds of The Poynter Institute, and Leo Bogart were sources and sounding boards. In North Carolina, conversations with Patricia Curtin, Frank Fee, Joe Bob Hester, Donald Shaw, Robert L. Stevenson and Xinshu Zhao kept the wheels turning. Lenny Lind and Kimberly Overcash-Clark of FGI oversaw the mail surveys. Librarians Barbara Semonche and Marion Paynter helped me discover, store and retrieve data in mass quantities.

Jane Cote, Bill Hawkins, Frank Hawkins, W. Davis "Buzz" Merritt and John Morton were seminar visitors while the project was unfolding. Present and former newspaper executives who shared their time and their viewpoints with Cote and me included Alvah Chapman, Larry Jinks, Tony Ridder, Doug McCorkindale, Al Neuharth, Erwin Potts and Gary Pruitt.

Bill Drewry, Lauren Rich Fine and John Morton helped me see the problem from the investor side. Thanks also to the people assembled by the Poynter Institute to ponder these problems, including Rick Edmonds, Steve Lacy, Geneva Overholser, Tom Rosenstiel, and Esther Thorson. William Friday provided wise career advice at a critical moment. Sue Quail Meyer made it all worthwhile.

And now the disclosures: I am a vested former employee of Knight Ridder, a recent consultant and contributor to Gannett's *USA TODAY*, and a sometime shareholder in Amazon.com, all of which receive mention in these pages. And, of course, I am still spending the fruits of Jack and Jim Knight's enterprise through the Knight Chair in Journalism that their foundation endowed. I hope they would be pleased. Neither the newspaper company nor the Foundation bears any responsibility for the message herein.

The more powerful conflict of interest is non-financial. I have been part of the newspaper business since the age of thirteen, when I delivered *The Clay Center* (Kansas) *Dispatch* on a seven-mile bicycle route through the poorest part of town. Newspaper people are my friends. The message in these pages is an attempt to warn and empower them. I pray that it helps.

Index

American Copy Editors Society (ACES), 156n16
American Journalism Review, 19, 64
American Meteorological Society, 243
American Newspaper Publishers Association, 124
American Press Institute, 224
American Press Managing Editors, 131
American Society of Newspaper Editors (ASNE): census of minorities in newsrooms by, 161, 191; and credibility issue, 18, 22, 30, 54, 56, 66, 89, 92; and ethical issues, 234, 240; focus groups for, 128; membership of, 78n15, 131–32; and quality issues, 131–32; and readership decline, 126, 143–44; and resignation of Jay Harris, 191; staffing surveys of, 102, 152, 160, 161, 162, 166; Statement of Principles by, 217; and Urban's research, 19
Analysts, 182, 185–86, 188–90, 193, 196–200, 210–11, 216
Anderson, Jack, 237
Angelo, Frank, 225
Annapolis Capital, 147
Annenberg, Walter, 180
Arant, Morgan David, 146, 148
Arizona Republic, 116, 120
Arkansas Gazette, 203
Arlington, Tex., 167
Art of Plain Talk (Flesch), 110
Art of Readable Writing (Flesch), 114
Associated Press (AP), 109, 110, 146, 179
Associated Press (AP) stylebook, 146, 148–49
Association for Education in Journalism and Mass Communication, 77
Athens, Ga., 57–58
Attention, scarcity of, 8–9
Auctions, online, 218
Audit Bureau of Circulations (ABC), 22–23, 22–23n32, 48, 50, 52, 79, 97–98, 123, 142n25, 156, 158, 161
Autocorrelation, 164–65
Average daily circulation, formula for, 5n2

Baldwin County, Ga., 50n7
Bar code scanners, 60

Barnes and Noble, 62
Baton Rouge Advocate, 116, 120, 138
Batten, James K., 68, 126, 185, 187, 188, 191, 192, 215
Believability, 18–19, 66, 67, 67n4, 69–70. *See also* Credibility
Bellows, Jim, 129
Bellows effect, 129
Bezanson, Randall, 216
Bibb County, Ga., 25
Biloxi Sun Herald, 52, 116, 120, 122, 135
Black, Cathleen, 43
Blair, Jayson, 63–64, 83–84, 171
Boca Raton, Fla., 21
Boca Raton News, 116, 119
Bogart, Leo, 78–79, 125–32, 143–44, 222
Boniche, Armando, 24n37
Borders, 62
Boston Globe, 4
Boulder County, Colo., 24, 24n35
Boulder *Daily Camera:* accuracy of reporting in, 90–92; advertising rates of, 52, 54, 56; circulation of, 24n35, 52; credibility of, 52, 54, 56, 95; editorial vigor of, 138; readability of, 117, 120; spelling and grammar errors in, 148n5, 149, 150, 153–55; staff of, 24n38
Bower, Joseph L., 221
Boyle, David, 1
Bradenton, Fla., 176
Bradenton Herald, 52, 116, 120
Brand identity, 43
Broward County, Fla., 23, 25, 27, 50n6, 70, 71, 94, 95, 95n14
Brown County, S.D., 24, 24n35, 26, 27
Budget: contingency budget, 38, 185; cost-cutting strategies for, 43; of newspapers, 36–37
Buffett, Warren, 13, 205
Bumba, Lincoln, 57
Bureau of the Census, 206
Business model of journalism: case of Thomson Newspapers, 212–15; community as market, 205–10; and competition from newer forms of media, 2, 217–18; and executive compensation, 215–17; factors in summarized,

About the Author

Photo by Dan Sears, University of North Carolina at Chapel Hill

PHILIP MEYER is Knight Chair and Professor of Journalism at the University of North Carolina at Chapel Hill. He is the author or coeditor of a number of books, including *Assessing Public Journalism* (University of Missouri Press) and *The Newspaper Survival Book: An Editor's Guide to Market Research.*

Printed in the United States
50199LVS00001B/19-66